MATH &
Mathematicians

M A T &H

Mathematicians:

The History of Math Discoveries Around the World

Volume 3

Leonard C. Bruno

Lawrence W. Baker, Editor

THOMSON
———★———™
GALE

Detroit • New York • San Diego • San Francisco • Cleveland • New Haven, Conn. • Waterville, Maine • London • Munich

THOMSON

GALE

Math and Mathematicians: The History of Math Discoveries Around the World, volume 3

Leonard C. Bruno

Project Editor
Lawrence W. Baker

Permissions
Kim Davis

Imaging and Multimedia
Leitha Etheridge-Sims, Robyn Young

Product Design
Cynthia Baldwin, Kate Scheible

Composition
Evi Seoud

Manufacturing
Rita Wimberley

ISBN 0-7876-6480-4
Printed in the United States of America
10 9 8 7 6 5 4 3 2 1

Contents

Contents

Entries by Mathematical Field

Algebra

Boldface type indicates volume number; regular type indicates page numbers.

Entries by Mathematical Field

Entries by
Mathematical Field

Group theory

Logic

Mathematical physics

Mathematics education

Entries by
Mathematical Field

Entries by
Mathematical Field

**Entries by
Mathematical Field**

Biographical Entries by Ethnicity

Boldface type indicates volume number; regular type indicates page numbers.

German

German American

Greek

Hungarian

Indian

Irish

Biographical
Entries by
Ethnicity

Reader's Guide

Mathematics has been described by one historian as "a vast adventure in ideas," but if that is so, we must always remind ourselves that it is individuals—real people—who have those ideas. However, unlike many other fields of science, mathematics seems to have only a few really well-known individuals whose names most people easily recognize. By high school, we all know something about the contributions of the mathematical greats like Euclid, Pythagoras, and Isaac Newton, yet the history of mathematics contains many more individuals whose accomplishments were nearly as important and whose lives may have been even more interesting. Volumes 1 and 2 of *Math and Mathematicians: The History of Math Discoveries Around the World* covered the early life, influences, and careers of fifty such individuals, telling the stories of those greats and near-greats whose contributions are not as well known as they should be.

In volume 3 of *Math and Mathematicians,* thirty-two additional individuals have been selected on the same basis as the first two volumes: to describe and explain their mathematical contributions but also to offer readers a sampling of how rich the history of mathematics is and how diverse are its contributors. The thirty

Reader's Guide

biographical entries in this volume (one entry, called "Thermometer Scale Originators," covers three men, Daniel Gabriel Fahrenheit, Anders Celsius, and William Thomson [Baron Kelvin of Largs]) include men and women from the ancient and modern world, who lived in nearly every major part of the globe. As with those individuals selected for the first two volumes, they too are linked by a common theme: that genius, hard work, determination, inspiration, and courage are multicultural, multiracial, and blind to gender differences.

Taken as a group, the thirty biographical entries in this volume include nine people who either were born or did their major work in the twentieth century, five of whom are still living. This surely tells us that not all the great mathematicians are found only in old history books. The oldest historical figure included here, an Egyptian scribe named Ahmes, is thought to have thrived around 1650 B.C.E., while the youngest living mathematician in this volume, Chinese-born American analyst Sun-Yung Alice Chang, was born in 1948. Ranging over the entire history of mathematics and making selections for only thirty entries for this volume obviously suggests that many more mathematicians were excluded than included, so it is not surprising that some truly deserving individuals were left out. However, with the help of an advisory board made up mostly of middle school and junior high librarians, we selected a solid group of mathematicians that spans the centuries.

In addition to the mathematical accomplishments that earned these individuals a high place in the history of mathematics, there are many fascinating personal stories that make them extremely interesting from a human perspective. While the number of child prodigies like János Bolyai, Arthur Cayley, Carl Jacobi, Blaise Pascal, and Ernest Wilkins Jr. is as high as might be expected, the accomplishments of two completely self-educated men, Joost Bürgi and Johann Lambert, are staggering. Two mathematicians were associated with significant "firsts": African American Elbert F. Cox, the first of his race to achieve a Ph.D. in mathematics, and American computer engineer J. Presper Eckert, who was responsible for producing the first fully electronic digital computer. Among some of the other more interesting or unique facts about this group, Englishman Thomas Harriot lived for a short time at what became the Lost Colony of Roanoke, while French geometer Gaspard Monge helped to create the metric system. Fourteenth-century French algebraist and geometer Nicole d'Oresme was also a Catholic bishop; the greatest geometer of modern times, Switzerland's Jakob Steiner, did not learn to read or write until he was fourteen; the person credited with writing the world's first computer program, Ada Lovelace, was a countess whose father was the famous English romantic poet known as Lord Byron; and the dashing and brilliant nineteenth-century Hungarian geometer János Bolyai could wield a sword as well as he could play a violin.

The individuals in this volume bridge a span of over thirty-six hundred years and show us that progress in mathematics, as with any science, begins with a single enquiring mind simply wanting to understand a little bit more about a subject.

Added features

Math and Mathematicians: The History of Math Discoveries Around the World includes a number of additional features that help to make the connection between math concepts and theories, the people who discovered and worked with them, and common uses of mathematics.

- Three tables of contents, alphabetically by mathematician, by mathematical field, and by mathematician's ethnicity, provide varied access to the entries. Entries from volumes 1 and 2 are included in the field and ethnicity tables of contents.

- A timeline provides a chronology of highlights in the history of mathematics.

- Dozens of photographs and illustrations bring to life the mathematicians, concepts, and ways in which mathematics is commonly used.

- Sidebars provide fascinating supplemental information about important mathematicians and theories.

- Extensive cross references make it easy to refer to other mathematicians and concepts covered in volumes 1 through 3; cross references to other entries are boldfaced upon the first mention in an entry.

- Sources for more information are found at the end of each entry so students know where to delve even deeper.

• A comprehensive cumulative index quickly points readers to the mathematicians, concepts, theories, and organizations mentioned in *Math and Mathematicians*. Entries from volumes 1 and 2 are included in the index.

Special Thanks

The author wishes to thank his wife, Jane, and his three children, Nat, Ben, and Nina, for again giving me their patience and understanding. None ever made me feel that this book was a rival or unwanted competitor for my time. In fact, all actually helped at one time or another, usually when I needed to understand something or have it explained to me in a simple, direct way.

Thanks also go to freelancers Mya Nelson, for her detailed copyediting; Leslie Joseph, for her eagle-eyed proofreading; and Theresa Murray, for her concise indexing. Much appreciation also goes to Marco Di Vita at the Graphix Group for his fine typesetting work.

At U•X•L, I must first thank senior market analyst Meggin Condino for again offering me the opportunity to work on another fine U•X•L project. My deepest debt, however, goes to senior editor Larry Baker who has the talent and humanity to be whatever the situation calls for. One of the pleasures of doing this book was renewing my long-distance friendship with this paragon of hard work and high standards. His sense of humor, total mastery of his craft, and easy understanding of a writer's situation make him able to handle virtually anything a writer can throw at him. Finally, he is not averse to simple hard work. I only wish we had more time to talk baseball.

Comments and suggestions

We welcome your comments on *Math and Mathematicians* as well as your suggestions for biographies to be featured in future volumes. Please write: Editors, *Math and Mathematicians*, U•X•L, 27500 Drake Rd., Farmington Hills, Michigan, 48331-3535; call toll-free: 1-800-347-4253; fax to 248-699-8097; or send e-mail via www.gale.com.

Advisory Board

Kari Deck
Librarian, Jim Hill Middle School, Minot, North Dakota

Jacquelyn Divers
Librarian, Cave Spring Middle School, Roanoke, Virginia

Elaine Ezell
Library Media Specialist, Bowling Green Junior High School, Bowling Green, Ohio

Marie-Claire Kelin
Library Media Services Teacher, Lincoln Middle School, Santa Monica, California

Eric Stromberg
Assistant Principal/Former Mathematics Teacher, Riley Middle School, Livonia, Michigan

Milestones in the History of Mathematics

50,000 B.C.E.   Primitive humans leave behind evidence of their ability to count. Paleolithic people in central Europe make notches on animal bones to tally.

c. 15,000 B.C.E.   Cave dwellers in the Middle East make notches on bones to keep count and possibly to track the lunar cycle.

c. 8000 B.C.E.   Clay tokens are used in Mesopotamia to record numbers of animals. This eventually develops into the first system of numeration.

3500 B.C.E.   The Egyptian number system reaches the point where they now can record numbers as large as necessary by introducing new symbols.

75,000 B.C.E.
Neanderthal man
can communicate
by speech.

38,000 B.C.E.
Homo sapiens
species evolves from
Neanderthal man.

12,000 B.C.E.
The dog is
domesticated from
the Asian wolf.

8000 B.C.E.
Earth's human
population soars to
5.3 million.

50,000 B.C.E.   40,000 B.C.E.   30,000 B.C.E.   20,000 B.C.E.   8000 B.C.E.

c. 2400 B.C.E. Mathematical tablets dated to this period are found at Ur, a city of ancient Sumer (present-day Iraq).

c. 2000 B.C.E. Babylonians and Egyptians use fractions as a way to help them tell time and measure angles.

c. 1800 B.C.E. The Babylonians know and use what is later called the Pythagorean theorem, but they do not yet have a proof for it.

c. 1650 B.C.E. The Rhind papyrus (also known as the Ahmes papyrus) is prepared by Egyptian scribe Ahmes, which contains solutions to simple equations. It becomes a primary source of knowledge about early Egyptian mathematics, describing their methods of multiplication, division, and algebra.

876 B.C.E. The first known reference to the usage of the symbol for zero is made in India.

c. 585 B.C.E. Greek geometer and philosopher Thales of Miletus converts Egyptian geometry into an abstract study. He removes mathematics from a sole consideration of practical problems and proves mathematical statements by a series of logical arguments. Doing this, Thales invents deductive mathematics.

c. 500 B.C.E. Greek geometer and philosopher Pythagoras of Samos formulates the idea that the entire universe rests on numbers and their relationships. He deduces that the square of the length of the hypotenuse of a right triangle is equal to the sum of the squares of the lengths of its sides. It becomes known as the Pythagorean theorem.

3500 B.C.E.
Human civilization begins as the Sumerian society emerges.

3000 B.C.E.
The Sahara Desert has its beginnings in North Africa.

2485 B.C.E.
The Great Sphinx carved from rock at Giza.

776 B.C.E.
First recorded Olympic games in Greece are held.

625 B.C.E.
Metal coins are introduced in Greece.

3500 B.C.E.   2750 B.C.E.   2000 B.C.E.   1250 B.C.E.   500 B.C.E.

c. 440 B.C.E.	Greek geometer Hippocrates of Chios writes *Elements of Geometry*, regarded by many as the first mathematical textbook.
c. 300 B.C.E.	Greek geometer Euclid of Alexandria writes a textbook on geometry called the *Elements*. It becomes the standard work on its subject for over 2,000 years.
c. 240 B.C.E.	Greek geometer Archimedes of Syracuse calculates the most accurate arithmetical value for pi (π) to date. He also uses a system for expressing large numbers that uses an exponential-like method. Archimedes also finds areas and volumes of special curved surfaces and solids.
c. 230 B.C.E.	Greek astronomer Eratosthenes develops a system for determining prime numbers that becomes known as the "sieve of Eratosthenes."
c. 100 B.C.E.	Negative numbers are used in China.
c. 150 C.E.	Greek geometer and astronomer Claudius Ptolemy's geometrical theories have important applications in astronomy.
c. 250	Greek algebraist Diophantus of Alexandria is the first Greek to write a significant work on algebra.
c. 320	Greek geometer Pappus of Alexandria summarizes in a book all acquired knowledge of Greek mathematics, making it the best source for Greek mathematics. French number theorist Pierre de Fermat later restores and studies Pappus's work.
c. 400	Greek geometer, astronomer, and philosopher Hypatia of Alexandria writes commentaries on

Milestones in the History of Mathematics

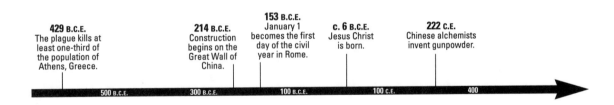

429 B.C.E. The plague kills at least one-third of the population of Athens, Greece.

214 B.C.E. Construction begins on the Great Wall of China.

153 B.C.E. January 1 becomes the first day of the civil year in Rome.

c. 6 B.C.E. Jesus Christ is born.

222 C.E. Chinese alchemists invent gunpowder.

500 B.C.E.   300 B.C.E.   100 B.C.E.   100 C.E.   400

Greek mathematicians Apollonius of Perga and Diophantus of Alexandria. She is the only woman scholar of ancient times and the first woman mentioned in the history of mathematics.

499 Hindu mathematician and astronomer Aryabhata the Elder describes the Indian numerical system. He also uses division to popularize a method for finding the greatest common divisor of two numbers.

700 Negative numbers are introduced by the Hindus to represent a negative balance.

820 Arab algebraist and astronomer al-Khwārizmī writes a mathematics book that introduces the Arabic word *al-jabr,* which becomes transliterated as algebra. His own name is distorted by translation into "algorism," which comes to mean the art of calculating or arithmetic. Al-Khwārizmī also uses Hindu numerals, including zero, and when his work is translated into Latin and published in the West, those numerals are called "Arabic numerals."

c. 825 Arab algebraist and astronomer al-Khwārizmī recommends the use of a decimal system.

1202 Italian number theorist Leonardo Pisano Fibonacci writes about the abacus, the use of zero and Arabic (Hindu) numerals, the importance of positional notations, and the merits of the decimal system.

1225 Italian number theorist Leonardo Pisano Fibonacci writes *Liber Quadratorum* in which he uses algebra based on the Arabic system.

629 The Koran is established as the holy book of Islam.

752 Japan's 55-foot Buddha statue is completed.

850 Coffee is discovered in East Africa.

1139 Civil war breaks out in England.

500 700 900 1100 1200

| 1299 | A law is passed in Florence, Italy, forbidding the use of Hindu-Arabic numbers by bankers. Authorities believe such numbers are more easily forged than Roman numerals. |

1364	French algebraist and geometer Nicole d'Oresme writes *Latitudes of Forms,* an important work on coordinate systems.
1482	The first printed edition of Greek geometer Euclid of Alexandria's geometry book, *Elements,* is published in Venice, Italy.
1489	The plus (+) and minus (–) symbols are first used in a book by German mathematician Johannes Widmann. They are not used as symbols of operation but merely to indicate excess and deficiency.
1535	Italian mathematician Niccolò Tartaglia demonstrates in a public forum his correct solution to the cubic equation. He later discloses in confidence his secret methods to another Italian mathematician, Girolamo Cardano, who later publishes the solution after learning that another Italian mathematician, Scipione dal Ferro, had discovered the solution as early as 1515. Cardano's paper correctly gives credit to both Tartaglia and dal Ferro.
1557	English mathematician Robert Recorde is the first to use the modern symbol for equality (=) in a book.
1570	The first complete English translation of *Elements,* by Greek geometer Euclid of Alexandria, appears.

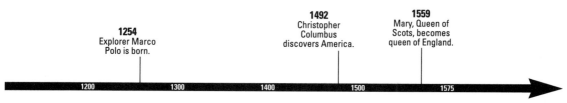

1254 Explorer Marco Polo is born.

1492 Christopher Columbus discovers America.

1559 Mary, Queen of Scots, becomes queen of England.

1200 1300 1400 1500 1575

1581	Italian mathematician Galileo discovers that the amount of time for a pendulum to swing back and forth is the same, regardless of the size of the arc. Dutch astronomer and mathematical physicist Christiaan Huygens would later use this principle to build the first pendulum clock.
1584	German algebraist Joost Bürgi begins work on improving a system of computing called "prosthaphairesis," a method of doing complicated multiplication by simple addition. By the end of the decade, he happened upon the idea of logarithms, and he created some actual conversion tables.
1585	Dutch mathematician Simon Stevin writes about the first comprehensive system of decimal fractions and their practical applications.
1591	French algebraist François Viète introduces the first systematic use of symbolic algebraic notation. He demonstrates the value of symbols by using the plus and minus signs for operations, vowels for unknown quantities (called variables), and consonants for known quantities (called parameters).
1594	Scottish mathematician John Napier first conceives of the notion of obtaining exponential expressions for various numbers, and begins work on the complicated formulas for what he eventually calls logarithms.
1601	English mathematician Thomas Harriot discovers the law of refraction of light.
1609	German astronomer and mathematician Johannes Kepler advances the development

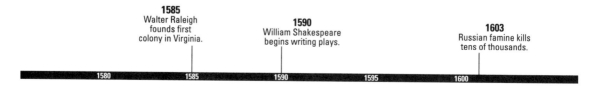

1585
Walter Raleigh
founds first
colony in Virginia.

1590
William Shakespeare
begins writing plays.

1603
Russian famine kills
tens of thousands.

1580   1585   1590   1595   1600

of the geometry of the ellipse as he attempts to prove that planets move in elliptical orbits.

1609   Italian mathematician Galileo improves upon the invention of the telescope by building a version with a magnification of about thirty times.

1614   Scottish mathematician John Napier invents "Napier's bones." This calculating machine consists of sticks with a multiplication table on the face of each stick. Calculations can be done by turning the rods by hand. He also publishes a book on logarithms.

1616   English mathematician Henry Briggs works with the Scottish inventor of logarithms, John Napier, to improve the base system of logarithms. Both agree that a base of 10 is the best method.

1619   German astronomer and mathematician Johannes Kepler shows that a planet's revolution is proportional to the cube of its average distance from the Sun.

c. 1621 English mathemetician William Oughtred invents the straight logarithmic slide rule.

1629   French number theorist Pierre de Fermat pioneers the application of algebra to geometry. Although French algebraist and philosopher René Descartes is credited with the invention and full development of analytic geometry, Fermat develops it earlier but does not publish his findings.

1631   English mathematician William Oughtred includes a large amount of mathematical symbol-

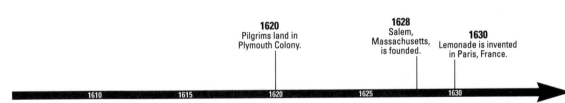

1620
Pilgrims land in Plymouth Colony.

1628
Salem, Massachusetts, is founded.

1630
Lemonade is invented in Paris, France.

1610   1615   1620   1625   1630

ism in a book he publishes, including the notation "×" for multiplication and "::" for proportion.

1632
Italian mathematician Galileo discounts the theory of an Earth-centered universe. As a result, the Roman Inquisition sentences him to life imprisonment.

1636
French number theorist Pierre de Fermat introduces the modern theory of numbers. His work includes theories on prime numbers.

1637
French algebraist and philosopher René Descartes introduces analytic geometry by demonstrating how geometric forms may be systematically studied by analytic or algebraical means. He is the first person to use the letters near the beginning of the alphabet for constants and those near the end for variables. He also includes a notation system for expressing exponents.

c. 1637
French number theorist Pierre de Fermat writes in the margin of a book a reference to what comes to be known as "Fermat's last theorem." This theorem remains the most famous unsolved problem in mathematics until it is solved in 1993 by Andrew J. Wiles. Fermat says he has a proof for the particular problem posed, but that the margin is too small to include it there.

1642
French geometer Blaise Pascal invents the first automatic calculator. It performs addition and subtraction by means of a set of wheels linked together by gears.

1644
French number theorist Marin Mersenne suggests a formula that will yield prime numbers.

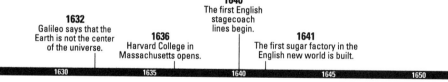

1632
Galileo says that the Earth is not the center of the universe.

1636
Harvard College in Massachusetts opens.

1640
The first English stagecoach lines begin.

1641
The first sugar factory in the English new world is built.

1630   1635   1640   1645   1650

These "Mersenne numbers" are not always correct, but they stimulate research into the theory of numbers.

1654   French number theorist Pierre de Fermat exchanges letters with French geometer Blaise Pascal in which they discuss the basic laws of probability and essentially found the theory of probability.

1657   Dutch astronomer and mathematical physicist Christiaan Huygens writes about probability.

1659   Swiss mathematician Johann Heinrich Rahn is the first to use today's division sign ($\div$) in a book. Later, English mathematician John Wallis adopts it and popularizes it through his works.

1660   English geometer Isaac Barrow begins his professorship at Cambridge University. Nine years later, he begins a professional relationship with English physicist Isaac Newton.

1662   English geometer William Brouncker becomes the first president of the prestigious Royal Society of London, a position he would hold until 1677.

1662   English statistician John Graunt is the first to apply mathematics to the integration of vital statistics. As the first to establish life expectancy and to publish a table of demographic data, Graunt is considered the founder of vital statistics.

1666   German logician Gottfried Leibniz begins the study of symbolic logic by calling for a "calculus of reasoning" in mathematics.

Milestones in the History of Mathematics

1667
The first recorded blood transfusion is performed.

1652
Capetown, South Africa, is founded.

1659
Typhoid fever is described for the first time.

1650   1653   1656   1660   1665

Milestones in the History of Mathematics

1668 — German mathematician and astronomer Nicolaus Mercator is the first to calculate the area under a curve using the newly developed analytical geometry.

1673 — German logician Gottfried Leibniz begins his development of differential and integral calculus independently of English physicist Isaac Newton.

1674 — Japanese mathematician Seki Kōwa publishes his only book, in which he solves 15 supposedly "unsolvable" problems.

1684 — German logician Gottfried Leibniz publishes an account of his discovery of calculus. He discovers it independently of English physicist Isaac Newton, although later than him. Newton, however, publishes his discovery after Leibniz in 1687. The timing of the discovery produces a feud between the two men.

1687 — English physicist Isaac Newton introduces the laws of motion and universal gravitation and his invention of calculus.

1690 — Massachusetts is the first colony to produce paper currency.

1693 — English astronomer Edmond Halley compiles the first set of detailed mortality tables, making use of statistics in the study of life and death.

1706 — English geometer William Jones is the first to use the sixteenth letter of the Greek alphabet, pi (π), as the symbol for the ratio of the circumference to the diameter of a circle.

1670 Minute hands appear on watches for the first time.

1682 Philadelphia, Pennsylvania, is founded.

1704 America's first regular newspaper begins publication.

1705 Thomas Newcomen invents the steam engine.

1670 1680 1690 1700 1710

1713	The first full-length treatment of the theory of probability appears in a work by Swiss mathematician Jakob Bernoulli.
1737	Swiss geometer and number theorist Leonhard Euler formally adopts the sixteenth letter of the Greek alphabet (π) as the symbol for the ratio of the circumference to the diameter of a circle. The ratio itself becomes known as pi. Following his adoption and use, it is generally accepted.
1748	Italian mathematician Maria Agnesi publishes *Analytical Institutions,* a large, two-volume work that surveys elementary and advanced mathematics. Agnesi is best known for her consideration of the cubic curve or what comes to be translated as the "witch of Agnesi."
1755	Nineteen-year-old French algebraist Joseph-Louis Lagrange sends a paper to Swiss geometer and number theorist Leonhard Euler concerning Lagrange's "calculus of variations." Euler is so impressed with the young man's work that he holds back his own writings on the subject, thus allowing Lagrange priority of publication.
1763	French geometer Gaspard Monge begins the study of descriptive geometry.
1767	Swiss-born German geometer Johann Lambert proves that the number for pi (π) is irrational.
1791	African American mathematician Benjamin Banneker assists in the surveying process of the new city of Washington, D.C.

1725
Antonio Vivaldi composes *The Four Seasons.*

1732
Benjamin Franklin revolutionizes the colonial postal service.

1714
Daniel Fahrenheit builds a mercury thermometer.

1754
Seven Years' War between the French and Indians begins.

1776
Declaration of Independence is written.

1715   1730   1745   1760   1775

1792	African American mathematician Benjamin Banneker publishes his first *Almanac*.
1792	The United States establishes its first monetary system, making the dollar its basic unit of currency.
1794	French geometer Gaspard Monge helps develop the metric system.
1795	France adopts the metric system.
1797	German mathematician Carl Friedrich Gauss gives the first wholly satisfactory proof of the fundamental theorem of algebra.
1813	English mathematician Charles Babbage cofounds The Analytical Society, whose general purpose is to revive mathematical analysis in England.
1816	French mathematician Sophie Germain receives an award for her paper on the mathematical theory of elasticity.
1820	English mathematician Charles Babbage conceives of the idea of calculation "by machinery." Over the next fifty years, he works on developing the "difference engine," but never succeeds. The technical requirements for such a machine turn out to be beyond the engineering ability of his time.
1821	French mathematician Augustin-Louis Cauchy publishes the first of three books on calculus.
1825	Norwegian mathematician Niels Abel first proves the impossibility of solving the general

1803
The United States nearly doubles, following the Louisiana Purchase.

1789
French Revolution begins.

1794
Eli Whitney invents the cotton gin.

1818
Russian socialist leader Karl Marx is born.

1780   1790   1800   1810   1820

quintic equation by means of radicals. This problem had puzzled mathematicians for two and a half centuries.

1829   Russian geometer Nicolay Lobachevsky describes his discovery of non-Euclidean geometry. This system includes the concept that an indefinite number of lines can be drawn in a plane parallel to a given line through a given point.

1829   German mathematical physicist Carl Jacobi publishes *Fundamenta nova theorae functionum ellipticarum,* which contains important work on the theory of elliptic functions.

1830   French algebraist and group theorist Évariste Galois is the first to use the word "group" in the technical sense and to apply groups of substitutions to the question of reducibility of algebraic equations.

1832   Hungarian geometer János Bolyai announces his discovery of non-Euclidean geometry, which he makes at about the same time as Russian geometer Nikolay Lobachevsky. His discovery is totally independent of Lobachevsky's, and when Bolyai finally sees Lobachevsky's work, he thinks it has been plagarized from his own.

1832   Swiss geometer Jakob Steiner lays the foundations for what would become known as projective geometry.

1833   Irish algebraist William Rowan Hamilton makes one of the first attempts at analyzing the basis of irrational numbers. His theory views

1834
The Braille system for the blind is invented.

1827
Contact lenses are invented.

1829
George Stephenson develops the railroad.

1822   1825   1828   1831   1835

both rational and irrational numbers as based on algebraic number couples.

1842   English applied mathematician Ada Lovelace translates an article about the computing machine ideas of English mathematician Charles Babbage. The notes she adds to the article produce the first clear mechanical explanation of Babbage's planned analytical engine and provide actual examples of how his machine might be instructed to perform certain tasks. These notes are now recognized as the world's first computer program.

1847   English logician George Boole maintains that the essential character of mathematics lies in its form rather than in its content. His work focuses on mathematics as symbolic rather than only "the science of measurement and number."

1854   English logician George Boole establishes both formal logic and Boolean algebra.

1854   German geometer Bernhard Riemann offers a global view of geometry. He develops further the ideas of Russian geometer Nikolay Lobachevsky and Hungarian geometer János Bolyai and introduces a new, non-Euclidean system of geometry.

1858   German number theorist Richard Dedekind conceives of the idea—later to be called the "Dedekind cut"—that treats the problem of irrational numbers in an entirely new manner, allowing irrational numbers to be categorized as fractions.

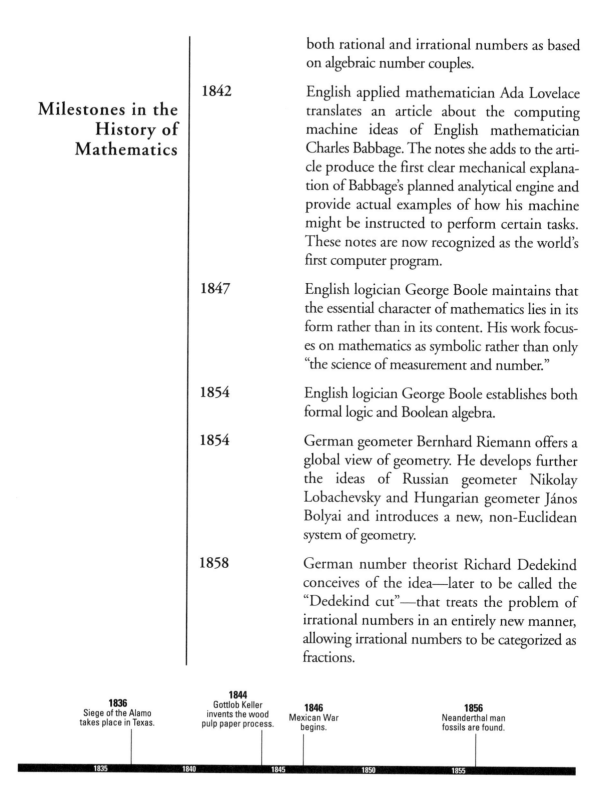

1836
Siege of the Alamo takes place in Texas.

1844
Gottlob Keller invents the wood pulp paper process.

1846
Mexican War begins.

1856
Neanderthal man fossils are found.

1835   1840   1845   1850   1855

1860	German geometer Bernhard Riemann uses the complex number theory to form the basis for most of the research in prime numbers for the next century.
1874	German mathematician Georg Cantor begins his revolutionary work on set theory and the theory of the infinite and creates a whole new field of mathematical research.
1874	Russian mathematician Sofya Kovalevskaya writes two papers on differential equations.
1883	English algebraist and geometer Arthur Cayley becomes president of the British Association for the Advancement of Science.
1884	Greenwich, England, is chosen as the site where the world's 24 time zones begin.
1888	Russian mathematician Sofya Kovalevskaya receives an award for her paper on the problem of how Saturn's rings rotate the planet.
1896	French analyst Jacques-Salomon Hadamard is the first to offer a correct proof showing that there is an infinite number of prime numbers.
1905	German American physicist and mathematician Albert Einstein writes five landmark papers that cover Brownian motion, the photoelectric effect, and his theory of relativity. It was with relativity that he devised his famous formula, $E = mc^2$.
1909	Danish mathematician Agner K. Erlang publishes "The Theory of Probabilities and Telephone Conversations." In this important work, he develops a formula that demonstrates the num-

Milestones in the History of Mathematics

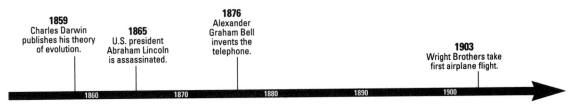

1859 Charles Darwin publishes his theory of evolution.

1865 U.S. president Abraham Lincoln is assassinated.

1876 Alexander Graham Bell invents the telephone.

1903 Wright Brothers take first airplane flight.

1860   1870   1880   1890   1900

ber of phone calls to arrive during a certain period of time, thereby allowing a phone company to calculate the fraction of callers who must wait when trying to place a call. This allows the company to provide more efficient service.

1913 Indian number theorist Srinivasa A. Ramanujan begins a five-year collaboration with English mathematician Godfrey Harold Hardy during which Ramanujan works on and solves many mathematical problems.

1921 German algebraist Emmy Noether publishes her studies on abstract rings and ideal theory which become important in the development of modern algebra.

1925 African American pure mathematician Elbert F. Cox is the first black to earn a Ph.D. in mathematics when he receives his degree from Cornell University.

1931 Austrian American mathematician Kurt Gödel publishes a paper whose incompleteness theorem startles the mathematical community. It states that within any rigidly logical mathematical system there are propositions that cannot be proved or disproved on the basis of the axioms within that system.

1933 Hungarian number theorist Paul Erdös discovers a proof for Chebyshev's theorem, which says that for each integer greater than one, there is always at least one prime number between it and its double.

1936 Chinese American geometrist Shiing-Shen Chern begins working with French number

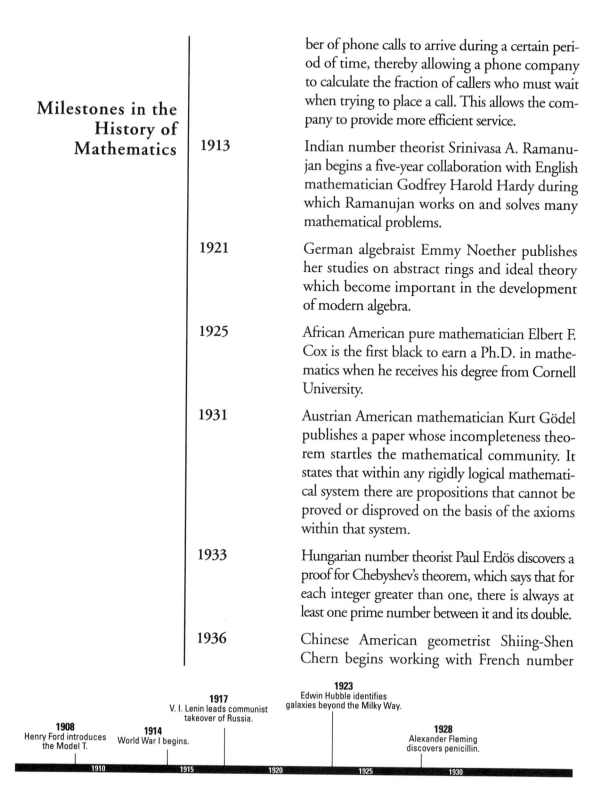

1908
Henry Ford introduces
the Model T.

1914
World War I begins.

1917
V. I. Lenin leads communist
takeover of Russia.

1923
Edwin Hubble identifies
galaxies beyond the Milky Way.

1928
Alexander Fleming
discovers penicillin.

1910   1915   1920   1925   1930

theorist Elie-Joseph Cartan on differential geometry.

1937   American mathematician Claude E. Shannon arrives at a connection between a computer's relay circuit and Boolean algebra.

1937   English algebraist and logician Alan Turing envisions an imaginary machine that would solve all computable problems and help prove the existence of undecidable mathematical statements.

1943   American computer scientist Grace Hopper joins the U.S. Navy, with whom she serves for 43 years.

1943   English algebraist and logician Alan Turing helps the World War II allies crack German codes.

1943   Harvard scientists, including American computer scientist Grace Hopper, build the Mark I, the world's first digital computer.

1944   Hungarian American number theorist John von Neumann and Austrian American economist Oskar Morgenstern develop a mathematical theory of games that comes to be known as game theory.

1944   African American mathematical physicist J. Ernest Wilkins Jr. begins a two-year stint at the University of Chicago, working on the Manhattan Project (the code name given to the American effort to build the first atomic bomb).

Milestones in the History of Mathematics

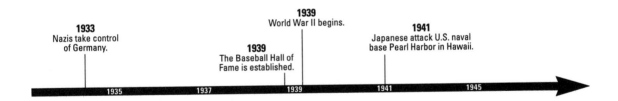

1933
Nazis take control of Germany.

1939
The Baseball Hall of Fame is established.

1939
World War II begins.

1941
Japanese attack U.S. naval base Pearl Harbor in Hawaii.

1935   1937   1939   1941   1945

| 1945 | Hungarian American number theorist John von Neumann presents the first description of the concept of a stored computer program. |

| 1946 | The ENIAC (Electronic Numerical Integrator and Computer), the first general-purpose electronic digital computer, is dedicated. Built by two Americans, computer engineer J. Presper Eckert and physicist and engineer John W. Mauchly, the ENIAC would influence all other electronic computers that would follow. Five years later, they completed the UNIVAC (Universal Automatic Computer), a computer for the everyday business world. |

| 1946 | The late Danish mathematician Agner K. Erlang is honored when the "Erlang unit" is adopted internationally to mean the total traffic volume of one hour. |

| 1947 | African American statistician David Blackwell describes "sufficiency," the process of simplifying a statistical problem by summarizing data. |

| 1948 | American logician Norbert Wiener produces a landmark paper that marks the beginning of cybernetics. |

| 1949 | University of Michigan students Evelyn Boyd Granville and Marjorie Lee Browne become the first African American women to receive Ph.D.'s in mathematics. |

| 1949 | American mathematician Claude E. Shannon formulates basic information theory, upon which much of today's computer and communications technology is based. |

1945
U.S. president Franklin Roosevelt dies during his fourth term.

1948
Jews in Palestine form the State of Israel.

1949
Mao Zedong becomes first leader of People's Republic of China.

1945   1946   1947   1948   1949

| 1951 | Fifteen nations found the International Mathematical Union to promote cooperation among the world's mathematicians and to more widely disseminate the results of mathematical research. |

| 1953 | American mathematician Claude E. Shannon publishes his pioneering work on artificial intelligence. |

| 1953 | African American mathematical physicist J. Ernest Wilkins Jr. and Herbert Goldstein publishes their work on the penetration of gamma rays, which is used in the design of nuclear reactors and radiation shielding. Their study on gamma ray penetration eventually becomes widely used for research in space and for other nuclear science projects. |

| 1956 | African American mathematician Evelyn Boyd Granville begins working at IBM as a computer programmer. |

| 1960 | The metric system is adopted by nearly every country in the world. |

| 1964 | American statistician and computer scientist Thomas E. Kurtz and Hungarian mathematician John George Kemeny develop a general-purpose computer language called BASIC, which soon becomes the most widely used language in the world. |

| 1966 | American algebraist Ruth Aaronson Bari receives her Ph.D. from Johns Hopkins University. The graph theory expert goes on to teach for over thirty years at George Washington Uni- |

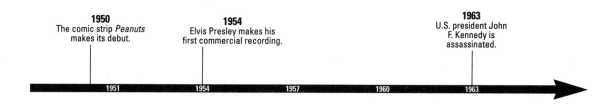

1950
The comic strip *Peanuts* makes its debut.

1954
Elvis Presley makes his first commercial recording.

1963
U.S. president John F. Kennedy is assassinated.

1951   1954   1957   1960   1963

Milestones in the History of Mathematics

versity, where her work with Ph.D. candidates earns her the nickname "doctoral mother."

1971 American mathematical statistician Mary Gray helps found and is the first president of the Association for Women in Mathematics.

1980 The work of English applied mathematician Ada Lovelace receives public recognition as the U.S. Department of Defense officially names a software language "Ada" in her honor.

1982 Polish-born Lithuanian mathematician Benoit B. Mandelbrot founds fractal geometry, a new branch of mathematics based on the study of the irregularities in nature.

1983 American logician and number theorist Julia Bowman Robinson becomes the first woman to be elected as president of the American Mathematical Society.

1986 Chinese-born American analyst Sun-Yung Alice Chang achieves national recognition when she deliver an address at the International Congress of Mathematicians.

1993 English-born mathematician Andrew J. Wiles announces his proof of "Fermat's last theorem." His 200-page paper is the result of a seven-year study on a problem left unsolved by French number theorist Pierre de Fermat 325 years earlier. Over the years, many mathematicians had declared it unsolvable.

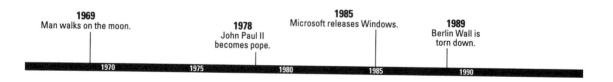

1969 Man walks on the moon.

1978 John Paul II becomes pope.

1985 Microsoft releases Windows.

1989 Berlin Wall is torn down.

1970   1975   1980   1985   1990

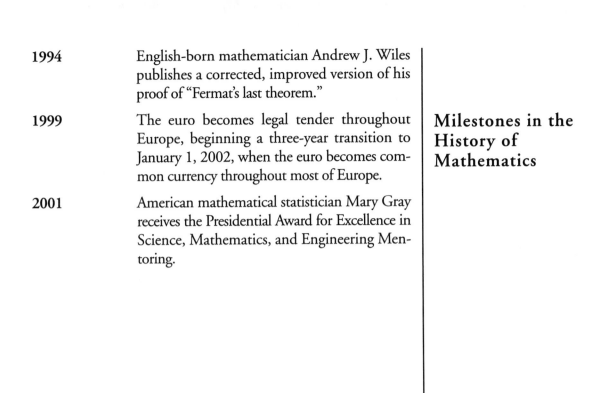

1994	English-born mathematician Andrew J. Wiles publishes a corrected, improved version of his proof of "Fermat's last theorem."
1999	The euro becomes legal tender throughout Europe, beginning a three-year transition to January 1, 2002, when the euro becomes common currency throughout most of Europe.
2001	American mathematical statistician Mary Gray receives the Presidential Award for Excellence in Science, Mathematics, and Engineering Mentoring.

Milestones in the History of Mathematics

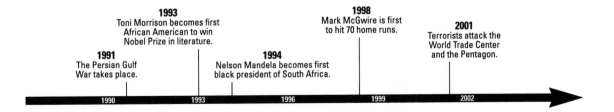

1993
Toni Morrison becomes first African American to win Nobel Prize in literature.

1998
Mark McGwire is first to hit 70 home runs.

2001
Terrorists attack the World Trade Center and the Pentagon.

1991
The Persian Gulf War takes place.

1994
Nelson Mandela becomes first black president of South Africa.

1990   1993   1996   1999   2002

Born c. 1680 B.C.E.
Egypt

Died c. 1620 B.C.E.
Egypt

Egyptian mathematics scribe

Ahmes

A hmes lived in Egypt nearly four thousand years ago and is associated with one of the oldest mathematical documents in existence. Called the Ahmes papyrus, or Rhind papyrus, this ancient document is scholars' main source of information about Egyptian mathematics.

Purchased in 1858

The Ahmes papyrus (pronounced AH-mace puh-PYE-russ) was bought at a resort on the Nile River in 1858 by a Scottish archaeologist named Alexander Henry Rhind. Papyrus was an early form of paper made in Egypt from the papyrus plant, a type of water grass that grows under the Nile River. The Egyptians made paper out of it by stripping thin coats from the stalk of the plant, spreading them out in a sheet, and placing another sheet of strips crosswise to form a thicker sheet that was then pressed and dried. Since this form of layered paper was made from a plant, it naturally decomposes and not a great deal of it has survived to modern times. Rhind had gone to Greece and Egypt for health reasons and eventually became interested in excavating and studying materials left behind from past human life and culture. In 1858, he saw an ancient piece of papyrus on sale and thought it might be both interesting and valuable.

By 1863, the papyrus that Rhind had bought was placed in the British Museum where scholars studied it. The papyrus was found to be a rare document that would prove especially important to the study of Egyptian mathematics. It was written in phonetic hieratic (pronounced hy-er-EH-tik) script. This style was a simplified version of pictorial hieroglyphics (pronounced hy-row-GLIH-fix), an everyday type of Egyptian writing that used pictures instead of letters to form words. The opening lines of the papyrus make it sound like a very important, almost secret kind of document since they are translated as "Directions for Obtaining the Knowledge of All Obscure Secrets." In fact, further translation shows that the complete document is really a collection of mathematical exercises designed mainly for students of mathematics.

The entire document is about 1 foot wide and 18 feet long, and contains about eighty-five problems that business and administrative clerks usually had to solve. For example, there are many bread and beer problems, such as dividing a certain number of loaves of bread among a certain number of people. The papyrus also has several problems that deal with measuring the slope of a pyramid. Although the papyrus is signed by a person who called himself Ahmes, historians know he was probably not the author of the document since he described himself as a scribe. In ancient Egypt, scribes belonged to a privileged class of educated people who served as teachers, editors, or copyists. As copyists, they would literally make copies of documents or texts. They also served as clerks or secretaries to Egyptian officials. Ahmes told nothing about himself except that he was a scribe. Historians have dated this particular papyrus to about 1650 B.C.E., although Ahmes wrote that the material contained in it came from an earlier version written during the Middle Kingdom, meaning it was written about 2000 B.C.E.

Egyptian mathematics

As a practical handbook that shows how to solve everyday problems, the Ahmes papyrus indicates how Egyptians counted and measured. The problems it illustrates show how Egyptians used

Ahmes

Egyptian hieroglyphics carved on a stone, similar to the style found on the Ahmes, or Rhind, papyrus.
Reproduced by permission of Archive Photos.

fractions (see entry in volume 1), and that they also had some form of **geometry** (see entry in volume 1). The papyrus also shows that Egyptians were able to measure **area** (see entry in volume 1) and **volume** (see entry in volume 2). Although the papyrus revealed a great deal, it does not actually give any rules for solving arithmetical problems. Some think that no rules were given because this document was prepared for advanced students who already knew the rules. Others think that it may simply have been the work of a student. Whatever the purpose, the

Ahmes

document demonstrates the Egyptian form of fractions called a two-term unit fraction. For them, all fractions were broken down into unit fractions, which had a numerator of two (and a different denominator). Scholars know how they used unit fractions but still do not know why they chose this form of fraction. The papyrus also reveals that Egyptians of this period never developed any real algebraic methods (easily solving for an unknown), and that overall, their methods for doing this were usually very long and complicated.

The Ahmes papyrus shows us that most Egyptian geometry was related to mensuration (pronounced men-suh-RAY-shun; the use of geometric methods to compute the length, area, or volume from given dimensions or angles). Thus, the Ahmes papyrus shows how to figure the area of certain shapes. The papyrus also indicates that Egyptian mathematics did not have anything that people today would recognize as formulas. Apparently, the student was given only the solution to a specific problem and was left on his or her own to see how it could be applied to other situations. Another piece of information the papyrus gives is that the Egyptians had a knowledge of geometric progression (in which the ratio of a term to its predecessor is always the same) and knew how to use it. For example, one problem in the Ahmes papyrus states only the following, "7 houses, 49 cats, 343 mice, 2,401 ears of spelt, 16,807 hekats." Most modern interpreters believe this to be a description of an exercise problem in which each of 7 houses contained 7 cats, and that each cat would eat 7 mice, and each mouse ate 7 ears of grain, with each ear of grain producing 7 measures of grain. This problem is similar to the familiar European nursery rhyme that goes like this: "As I was going to St. Ives, I met a man with seven wives. Every wife had seven sacks, every sack had seven cats, every cat had seven kits. Kits, cats, sacks, and wives, how many were going to St. Ives?"

Finally, the Ahmes papyrus about Egyptian mathematics shows that there was only a minimal use of mathematical symbols. For example, **addition** (see entry in volume 1) and **subtraction** (see entry in volume 2) were represented by the legs of a man coming (addition) and going (subtraction).

In short, the Ahmes or Rhind papyrus has been a key tool in furthering scholars' understanding of ancient Egyptian mathematics.

Arabs Preserve Greek Mathematics

Hundreds of years after the Ahmes papyrus was created, providing much information about Egyptian mathematics to the academic world in the nineteenth century, the work of Greek mathematicians was noticed by Islamic scholars. With the fifth century collapse of the Roman Empire in the West, the Middle Ages—a time when learning and knowledge was misplaced and nearly forgotten—are said to have begun. For centuries after, few in the West could read or write, let alone understand mathematics. However, while the ancient Greek tradition of mathematics had been all but lost to the West, it was being cared for and added to in the East by Islamic scholars who wrote in Arabic.

The Greek tradition of mathematics was looked after by a school of Arabic scholars who translated the Greek classics of **Apollonius of Perga** (c. 250 B.C.E.–c. 180 B.C.E.), **Archimedes of Syracuse** (287 B.C.E.–212 B.C.E.; see entry in volume 1), **Euclid of Alexandria** (c. 325 B.C.E.–c. 270 B.C.E.; see entry in volume 1), and **Claudius Ptolemy** (c. 100–c. 170), and preserved many a Greek classic that would otherwise have been lost. This mathematical knowledge was essentially lost to the West until the late eleventh century when the Spanish city of Toledo was taken back from the Moors and Western scholars flocked there to learn the science of the Arabs. Thus began the great wave of translations, in which many of Toledo's flourishing Jewish population who were fluent in Arabic, first translated the Arabic into Spanish, followed by a Christian scholar who would translate the Spanish into Latin. Many think the greatest of all these translators was Gerard of Cremona (1114–1187), an Italian living in Toledo, who is credited with translating more than eighty works. It is only because of the existence of these Arabic texts and the exchange between the three major civilizations of the Mediterranean—Islamic, Jewish, and Christian—that the mathematics of the Greeks was preserved and became available to a wider audience.

These Arab texts exerted yet another lasting influence. Many of their names and terms were Latinized and became part of the mathematical language. For example, the word "algorithm" came from the term *Algorithmus* which was a Latinization of the name of Arab algebraist and astronomer **al-Khwārizmī** (c. 780–c. 850; see entry in volume 2). The English word "algebra" came from the Arabic word *al-jabir,* which was a word in the title of a mathematical book by al-Khwārizmī. Many other familiar symbols and words are Arabic in origin, such as "sine" which comes from the Latin translation ("sinus") of an Arabic word. The familiar term "digit" was actually the smallest Egyptian unit, being equal to the width of one finger. Finally, the well known "Arabic numerals" of the West should in fact be properly called Hindu-Arabic numerals since they were invented by the Hindus of northern India and then adopted by the Arabs and transmitted to the West.

For More Information

Allen, Don. *The Ahmes Papyrus.* http://www.math.tamu.edu/~don.allen/history/egypt/node3.html (accessed June 10, 2002).

Burton, David M. *The History of Mathematics: An Introduction.* New York: McGraw-Hill, 1999.

Ahmes

Obretenov, Christos. "Ahmes: What Were You Thinking?" *History of Mathematics, Simon Fraser University.* http://www.math.sfu.ca/History_of_Math/Egypt/Ahmes/main.html (accessed June 10, 2002).

O'Connor, J. J., and E. F. Robertson. "Ahmes." *The MacTutor History of Mathematics Archive.* School of Mathematics and Statistics, University of St. Andrews, Scotland. http://www.groups.dcs.st-andrews.ac.uk/~history/Mathematicians/Ahmes.html (accessed June 10, 2002).

Born November 17, 1917
Brooklyn, New York

American algebraist

Ruth Aaronson Bari

The life and professional career of Ruth Aaronson Bari are examples of patience, determination, and enthusiasm. Despite her obvious mathematical ability and love of mathematics, World War II (1939–45) and family responsibilities caused Bari to take a break from her field for almost two decades. However, after raising a family and proving her ability a second time, she obtained her Ph.D. and became an expert in a highly complex and specialized field of mathematics known as graph theory.

Mathematics comes easy

Ruth Aaronson Bari is the youngest daughter of Polish immigrants. Her father, Israel Aaronson, who was a grocer, and her mother, Becky Gursky, who also worked in the family's stores, had one other daughter named Ethel. Bari went to several different public schools in Brooklyn, New York, and graduated from the all-girls Bay Ridge High School. Although mathematics always came very easy to Bari, many in her high school did not think mathematics was of much importance in the education of young women. Besides her own interest in mathematics, Bari's father worked with her in reasoning out some of her **algebra** (see entry in volume 1) problems, so she felt encouraged at home. As Bari progressed rapidly, she felt that she should take advanced algebra courses. However, her high school teacher offered her only a book on higher algebra and told her to read it on her own. Not at all discouraged, Bari did just that and went on to earn a medal for her mathematics work when she graduated. She then took the New York Regents exam, a very difficult test for high school graduates that separated the truly talented from the average students, and achieved a nearly perfect score.

Begins her serious study of mathematics

Although Bari showed exceptional ability at an early age and was mostly self-taught in mathematics, she had no idea what the field of mathematics was all about. In fact, while in high school, she actually believed that mathematics ended with calculus (the field of mathematics that deals with rates of change and motion). Fortunately for her, she applied to Brooklyn College and was accepted. That school was unusual in that it was a small but very selective institution that did not charge tuition and even provided free books. There, Bari naturally continued to study mathematics and took a course in abstract algebra. In that course, she found her interest in mathematics deepen, and she decided that she would become a mathematician, even though it was a subject that did not often lead to a good, steady job.

Bari was in college during the mid-1930s, during the middle years of the Great Depression (1929–41). This was a time of severe economic troubles for the United States and the world, and jobs were scarce. Therefore, as a sort of insurance policy, Bari also studied accounting in case she needed to find a bookkeeping job when she graduated. In 1939, Bari graduated with a bachelor's degree in mathematics and planned to go to graduate school if she could. Her graduation summer was spent working on Staten Island, and it was there that she met Arthur Bari, a diamond setter who had at one time studied engineering. Plans for school were soon replaced by plans for marriage, and the two were married on November 22, 1940. Bari had already taken and passed a civil service entrance examination, and the couple moved to Washington, D.C., where she obtained a job as a statistical clerk for the 1940 census.

Bari still longed to continue her mathematical education, and since her husband had joined the Marines and was sent to the

Ruth
Aaronson Bari

South Pacific after the United States entered World War II in 1941, she enrolled at Johns Hopkins University. She attended courses at night and worked during the day. By 1943, she obtained her master's degree, and was encouraged by the department chairman to go on for her Ph.D. Bari had no sooner begun to work toward this degree when World War II ended and American soldiers began streaming back home. The university suddenly found that it could no longer offer her, or any other women, fellowships or even teaching jobs, since these were now reserved for men. All American universities felt obligated to help these returning men continue their education, and Johns Hopkins was no exception. Bari left school with only a master's degree and went to work for Bell Telephone Laboratories in New York City, before she was hired as an instructor at the University of Maryland.

From motherhood to Ph.D.

At the University of Maryland, Bari's work was again interrupted when she became pregnant. In 1948, her first daughter, Gina, was born and she resigned her position at Maryland. The birth of her first child began a new, decidedly non-mathematical, phase in her life as she decided to stay at home and raise her child. She would have two more daughters, Judi, born in 1949, and Martha, born in 1951. It was not until her youngest daughter was four years old and able to attend school that Bari decided to return to school herself. With the support and encouragement of her husband, she became determined to fulfill her dream of attaining a Ph.D. in mathematics and applied for readmission to Johns Hopkins in 1955. She was admitted, but only on the condition that she redo all of the course work she had already completed years before for a master's degree. This was required not only to prove that she could still do mathematics, but to be sure that she had all of the preparation required for doctoral-level research.

Glad to be back at school, Bari had to balance being a mother to three children under ten, a doctoral candidate, and an instructor. While a student at Johns Hopkins, she had become a full-time instructor at Maryland because an instructor's pay was better than that of a teaching assistant at Johns Hopkins. She worked late at night while her family slept and no one would interrupt her. Eventually, her steady, hard work paid off as she completed her disserta-

**Ruth
Aaronson Bari**

Ruth Aaronson Bari

tion and received her Ph.D. in 1966. She was forty-seven years old. When she applied for a teaching position at Washington-area universities, she received job offers from nearly every one. She chose George Washington University where she became a full professor and where she remained until she retired at the mandatory age of seventy. Bari's work with Ph.D. candidates at George Washington earned her the affectionate title of "doctoral mother" due to her patience and encouragement. She is also recognized for her influential work in the complex specialty of graph theory known as chromatic polynomials. Chromatic means something relating to color, and a polynomial is a form of algebraic equation. Bari's mathematical work on this part of graph theory helps mapmakers determine the number of ways to best color a map (without the same colors touching) depending on how many colors are available.

Bari's obvious joy and pleasure at finally pursuing the mathematics she so loved was apparent to everyone who knew or worked with her. She sincerely loved the academic life and said that she never got bored teaching her specialty over and over, since every class contained a different set of students. In a brief essay about her mother, Gina Kolata, Bari's oldest daughter, who is a science journalist for the *New York Times,* said in *Biographies of Women Mathematicians* that her mother always counseled her children to follow their heart. She spent years "encouraging us to find a subject that enthralled us and to pursue it, paying no heed to whether there would be a job for us at the end of our studies." She is also recognized as a woman who knew what she wanted and followed her heart.

For More Information

Fasanelli, Florence D. "Ruth Aaronson Bari." In *Women of Mathematics: A Bibliographic Sourcebook.* Edited by Louise S. Grinstein and Paul J. Campbell. Westport, CT: Greenwood Press, 1987, pp. 13–16.

Kolata, Gina. "Ruth Aaronson Bari." *Biographies of Women Mathematicians.* Agnes Scott College. http://www.agnesscott.edu/lriddle/women/bariruth.htm (accessed June 10, 2002).

Born October 1630
London, England

Died May 4, 1677
Cambridge, England

English geometer and theologian

Isaac Barrow

Isaac Barrow is best known today for his connection to the great English mathematician and physicist **Isaac Newton** (1642–1727; see entry in volume 2). As a seventeenth century mathematician, Barrow was considered by the English as second only to Newton. He published several excellent books on **geometry** (see entry in volume 1) and also is noted for his contributions to the field of optics (the study of light).

Isaac Barrow.
Courtesy of the Library of Congress.

Starts as a bully in school

Isaac Barrow's father, Thomas Barrow, was a prosperous dealer in cloth who also had connections to the court of King Charles I (1600–1649). Isaac's mother, Anne Buggin, died when he was four, and he was eventually sent by his father to Charterhouse, a school noted for its emphasis on a classical education. Isaac lived at home while he went to school, and gave his father so much trouble that he is said to have prayed that if God were to ever take one of his children he could best spare young Isaac. At school, Isaac earned a reputation as a rebel and a bully, and seemed more interested in fighting other boys than learning or doing his schoolwork. His father soon sent him off to Felstead, a school in Essex

Isaac Barrow

whose headmaster (principal) was known for both his strict discipline and for his ability to change a boy's bad habits and make him a good student. Barrow spent four years there and, remarkably, grew into an excellent student who studied Latin, Greek, Hebrew, French, logic, rhetoric (pronounced REH-tuh-rik; the art of speaking and writing) and the classics.

Barrow did extremely well at Felstead, but when his father suffered business losses after a rebellion destroyed the cloth trade in Ireland, he could not pay for school. The school headmaster recognized Barrow's potential and allowed him to pay his tuition by tutoring. In 1646, Barrow enrolled at Trinity College in Cambridge, and was allowed to pay his tuition by doing work that servants usually did. He was able to learn some mathematics there, but he did not study it in depth until after he graduated. By 1649, at the age of nineteen, Barrow received his bachelor's degree from Trinity and received a scholarship to continue. Three years later, he had earned his master's degree at Trinity, and this allowed him to become a college lecturer and university examiner there. At Trinity, Barrow had made a promise that he would study divinity or religion. He eventually began to study astronomy, which, in turn, led him to study geometry.

Loyalty to the king earns him trouble

The mid-1600s was an especially troubled time in England as that country was experiencing a civil war between those who supported King Charles's rule and those who wanted to overthrow the king and to give power to the people's representatives known as the Parliament (pronounced PAR–luh–munt; the English equivalent of the U.S. Congress). By 1646, the king's forces had lost, and although Charles tried to escape, he was captured and put on trial in 1648. A year later, Charles was beheaded and the English kingdom appeared dead with him. All of England was split between those who had been the king's supporters and those who were not. Barrow, like his father, was very much on the side of the king and did not hesitate to let people know this. These views made him many enemies and were probably the reason that he was not appointed professor at Cambridge University, a position he had earned.

In 1655, Barrow took advantage of an opportunity to leave England and study abroad, and was gone from his homeland for nearly five years. Having won a scholarship, he sold his private collection of books and traveled first to Paris, France. However, he found that there were few mathematicians in France from whom he could learn anything, so he went to Florence, Italy, after ten months. In Florence, he met more capable men, especially Italian mathematician Vincenzo Viviani (1622–1703). Viviani had been the last pupil of the great Italian mathematician and physicist **Galileo** (1564–1642; see entry in volume 1). After his time in Italy, Barrow sailed for Turkey. On this journey, his ship was attacked by pirates; reportedly, Barrow's braveness and courage saved the ship from being captured. Barrow returned to England in late 1659 after the king's rule had been restored and a more friendly attitude was waiting for him.

Becomes an established figure

In 1660, Barrow was given the professorship at Cambridge that had been denied him, and he also decided it was time to be ordained in the Anglican Church. At the same time he became a religious official in that church, he accepted another professorship in order to make up for his modest salary at Cambridge. He became professor of geometry at Gresham College and also taught astronomy there. By 1663, Barrow had established himself as a major figure in England, and in that year Cambridge asked him to become the first person to hold a new position called the Lucasian professor of mathematics. In 1663, Barrow also became one of the first scientists to be elected to a newly formed scientific society called the Royal Society of London for the Improvement of Natural Knowledge. This is the oldest scientific society in England and is one of the oldest in Europe. Barrow taught at Cambridge until 1669 when he gave up his Lucasian professorship so that Newton could have it. It is said that after he met Newton and saw his towering genius, Barrow did no more work in mathematics. After stepping down in favor of Newton, Barrow was appointed royal chaplain to King Charles II (1630–1685). Now devoting all his time and energy to religion, Barrow returned to Trinity College in 1673.

Barrow died in 1677 soon after he became feverish while traveling to London. He died while trying to cure himself by fasting and taking opium (pronounced OH-pee-uhm). Opium is a very powerful drug made from the fruit of the opium poppy, and can cause death if taken in too large a dose. Barrow had once tried this same

Isaac Barrow

Isaac Newton (above) and Isaac Barrow were considered the top two mathematicians in England during the 1600s. *Reproduced by permission of Archive Photos.*

drug when he was sick during his travels in Turkey and he believed that it cured him. This time, however, he died of a drug overdose.

Connection with Isaac Newton

During his lifetime, Barrow was considered second only to Newton as a mathematician in England. Newton was twelve years younger, and although there is no real evidence that he was a student of Barrow's, many argue that Barrow did serve as a mentor for

Newton. Others say that he attended some of Barrow's lectures on optics and therefore became interested in that field himself. It is known for certain, however, that in late 1669 they did have some sort of working relationship, especially since Barrow resigned his professorship that year so that Newton could have the post.

Publishes his mathematical lectures

Barrow's best mathematical work may have been his geometrical and mathematical lectures. His *Lectiones mathematicae* (Mathematical Lectures) were published after his death and discuss the foundations of mathematics. In these lectures, however, Barrow refused to consider **algebra** (see entry in volume 1) as a field or branch of mathematics, and said instead that it was simply a sort of logical tool. Although his *Lectiones geometricae* (Geometrical Lectures) did not contain any really original material, they brilliantly discussed the work of some of the top mathematicians in Europe and made their work understandable to a new generation of scholars. Barrow's first publication, *Euclidis Elementorum* (The Elements of Euclid), was a popular undergraduate text and was even published in a pocket-sized edition.

Barrow never married, and was described as slim, strong, and athletic. He was known as a sloppy dresser, a man with a very good sense of humor, and a heavy smoker. His lectures on religion were as well received as those on mathematics.

For More Information

Abbott, David. *Mathematicians.* New York: Peter Bedrick Books, 1985.

Feingold, Mordechai. *Before Newton: The Life and Times of Isaac Barrow.* New York: Cambridge University Press, 1990.

O'Connor, J. J., and E. F. Robertson. "Isaac Barrow." *The Mac-Tutor History of Mathematics Archive.* School of Mathematics and Statistics, University of St. Andrews, Scotland. www.groups.dcs.st-andrews.ac.uk/~history/Mathematicians/Barrow.html (accessed June 18, 2002).

Whiteside, D. T. "Isaac Barrow." In *Biographical Dictionary of Mathematicians.* New York: Charles Scribner's Sons, 1991, pp. 180–83.

Isaac Barrow

Wilkins, D. R. "Isaac Barrow (1630–1677)." *A Short Account of the History of Mathematics.* School of Mathematics, Trinity College, Dublin. http://www.maths.tcd.ie/pub/HistMath/People/Barrow/RouseBall/RB_Barrow.html (accessed June 18, 2002).

Born December 15, 1802
Kolosvar, Hungary (now Cluj, Romania)

Died January 27, 1860
Maros-Vásárhely, Hungary (now Tîrgu-Mureş, Romania)

Hungarian geometer

János Bolyai

János Bolyai is now recognized as one of the fathers of modern mathematics, despite the fact that he was mostly unknown in his lifetime. He is credited, along with Russian geometer **Nikolay Lobachevsky** (1792–1856; see entry in volume 2), with the independent invention of an entirely new kind of **geometry** (see entry in volume 1). The usefulness of this new geometry would not become apparent until many decades passed and it was finally applied to modern physics by German American physicist and mathematician **Albert Einstein** (1879–1955; see entry in volume 1).

Mathematics and music

János Bolyai (pronounced BOWL-yoy) was a Hungarian mathematician, although he was born in a part of eastern Europe that is now in Romania. The son of Farkas Wolfgang Bolyai and Susanna Benko Arko, Bolyai showed a natural ability for both mathematics and music at an early age. There is no doubt that he was a mathematical prodigy (pronounced PRAH-dih-gee; a child with extraordinary talent), since he understood functions of trigonometry—such as sine, cosine, and tangent—at the age of five. Trigonometry is the study of the properties of **triangles** (see entry in volume 2), and is not an easy subject. Bolyai also demonstrated remarkable musical ability and was considered a highly skilled violinist.

János Bolyai

Bolyai's father was a teacher of mathematics, physics, and chemistry at the college level, and was responsible for János's early education. When his father was a student of mathematics and went to study in Göttingen, Germany, he became friends with German mathematician **Carl Friedrich Gauss** (1777–1855; see entry in volume 1). Gauss (pronunciation rhymes with HOUSE) himself was a prodigy who came to be regarded as one of the three greatest mathematicians of all time. The older Bolyai wanted his gifted son to study with the great Gauss. First, however, he entered thirteen-year-old János into the Evangelical Reformed College where he taught. At thirteen, Bolyai already had mastered calculus, and during his five years at the university he quickly mastered the physical sciences as well as mathematics.

Mathematics and swords

Upon leaving Evangelical, János disappointed his father by enrolling in the imperial engineering academy in Vienna, Austria, instead of going to Germany to study with Gauss. Bolyai remained at that academy until 1822 and received a military education and graduated with a commission. While at school, he also pursued his own deep interest in mathematics. However, Bolyai was not the quiet, studious type at the academy, and it was during these years that he became an expert swordsman. As a dashing, slim, and very athletic young man, Bolyai soon showed everyone he could handle a sword as well as a violin. The flamboyant Bolyai would accept a dueling challenge from anyone, sometimes defeating several opponents in a row. It is said that one time he crossed swords with thirteen consecutive opponents and defeated them all. To add to his growing reputation, Bolyai asked only that he be allowed to pause in between matches to play his violin. Despite his many adventures at the academy, he nonetheless always found time to pursue a mathematical challenge.

"For God's sake, I beseech you, give it up"

While still a student at the academy, Bolyai began to search for and consider mathematical concepts that would both challenge him and serve as a contribution to the field he loved. It did not take him long to settle on something so challenging that his father had long before given it up as hopeless. His father had labored long and hard trying to prove the Euclidean (pronounced yoo-CLUD-ee-in) theory of parallels. This principle states that through any given point, one and only one line can be drawn, infinitely, and in both directions. In other words, parallel lines never meet. This was only one of several postulates or axioms (facts widely accepted as true) that were put forth centuries ago by Greek geometer and logician **Euclid of Alexandria** (c. 325–c. 270 B.C.E.; see entry in volume 1). However, this axiom had long bothered many mathematicians, since it was the only postulate that was not "self-evident," or obviously true. Although few really doubted its truth, there was, in fact, no real way to verify or prove it by experience, since it dealt with a line being drawn to infinity.

Although Bolyai's father had given up trying to prove this postulate, it still must have occupied his thoughts, for his son soon developed a similar fascination for it. The elder Bolyai worried that his son would work at it extensively yet be defeated by it as he had, and he wrote to him, "For God's sake, I beseech you, give it up. Fear it no less than sensual passions because it, too, may take all your time, and deprive you of your health, peace of mind, and happiness in life." This is an amazing thing for a father to say to his son, and it shows how deeply the older man had been possessed by this problem. As might be expected of the younger Bolyai, who had gone against his father's wishes and not studied with Gauss, he ignored his father's warnings and pushed on to try and prove this famous postulate.

"I have created another entirely new world"
It was around 1820, while he was still at the military academy, that Bolyai began to think that perhaps a proof of the Euclidean theory of parallels was not possible. This led the brilliant young mathematician to think about it in a radically different way from what most people had thought. What if he actually tried denying the postulate and went from there? Bolyai did just that, and began to build a new geometry that did not depend on Euclid's parallel postulate. Over the next three years, he worked on what he called his theory of absolute space in which he showed that more than one parallel line could be drawn through a point. When he finished, Bolyai realized that his results contained no contradictions and

János Bolyai

were as self-evident as Euclid's. What he did was invent an entirely different geometry that was as valid or true as Euclid's. The real revolutionary implication of this, however, is that it showed for the first time that pure or theoretical mathematics need not be true or false in the same way as physics. Rather, it need only be self-consistent. Bolyai thus invented what mathematicians now call non-Euclidean geometry and laid the foundations for mind-stretching ideas that made modern mathematics truly modern. It is safe to say that without non-Euclidean geometry, Einstein would not have been able to develop his general theory of relativity—a theory that transformed twentieth century physics.

In 1825, Bolyai visited his father and handed him a manuscript of a paper titled "The Absolute True Science of Space." His father could not see the truly revolutionary aspects of his son's discovery and essentially rejected it. However, as a good father, he sent the manuscript to Gauss. What happened next would change the lives of both Bolyais forever. To their surprise and shock, Gauss wrote back that he himself had discovered the very same geometry some thirty to thirty-five years before, and that "now I have been saved the trouble [of writing a paper]." Some scholars doubt that

Gauss was being truthful, although many more are of the opinion that he deliberately hid his discovery out of a lack of courage. In fact, Gauss had an enormous reputation, which may have been jeopardized or put in danger if he had published such a strange and radically different concept.

From this point on, both father and son met nothing but disappointment and misfortune. Young Bolyai was especially hurt and humiliated. The great Gauss had made his judgment, saying basically that he had discovered the same thing years before, and therefore it was nothing so special. What made matters worse was that Gauss remained silent about Bolyai's discovery and informed no one. Today, scholars know of Bolyai's discovery because in 1832–33, his father published his son's paper as an addendum or appendix to his own two-volume work on elementary mathematics. As if Gauss's dismissal of Bolyai's geometry was not bad enough, poor Bolyai received yet another shock in 1832 when he read a paper written in 1829 by an unknown Russian named Nikolay Lobachevsky. The paper described a non-Euclidean geometry so like his own that Bolyai thought at first Lobachevsky had copied it from his own work. Lobachevsky would have his own trouble getting his discovery recognized for what it was, but he had at least published his paper before Bolyai and would get credit for the discovery (even though Bolyai had done his work by 1825).

A sad end

Both father and son reacted to this fate with anger and bitterness. Bolyai gave up his military commission and mathematics as well. Although he left behind a mass of mathematical manuscripts, he published nothing further in his lifetime. After his mother died, he moved into his father's home in 1833, but the two could not get along peacefully. Bolyai eventually moved to a nearby village where he lived with a woman named Rosalie von Orban, and had three children. In 1837, father and son gave mathematics one last try for glory and submitted their work in hopes of winning an important mathematical prize. Once again, they met with disappointment as their submission was too complicated or involved for the judges to understand. Bolyai's father died in 1856, and that same year Bolyai left von Orban and his children to live alone at his family's estate. During his military service, Bolyai had regularly suffered from chronic fevers, and as he lived alone his health got worse. In 1860, he died after a lengthy illness, unrecognized and unknown for the mathematical genius that he was.

János Bolyai

For More Information

Abbott, David, ed. *Mathematicians.* New York: Peter Bedrick Books, 1985.

"Bolyai, János or Johann (1802–1860) and Lobachevsky, N. I. (1793–1856)." *Thinkquest.* http://library.thinkquest.org/22584/temh3019.htm (accessed June 18, 2002).

Franceschetti, Donald R., ed. *Biographical Encyclopedia of Mathematicians.* New York: Marshall Cavendish, 1999.

Struik, D. J. "János Bolyai." *Biographical Dictionary of Mathematicians.* New York: Charles Scribner's Sons, 1991, pp. 296–97.

Born February 1561
Warleywood, Yorkshire, England

Died January 26, 1630
Oxford, England

English geometer and educator

Henry Briggs

Henry Briggs is the person most responsible for the general acceptance of **logarithms** (see entry in volume 2). The early seventeenth century invention of logarithms was a major calculating breakthrough, since it reduced the tedious **multiplication** (see entry in volume 2) of large numbers to the simple and quick **addition** (see entry in volume 1) of smaller ones. It was Briggs who improved this new invention and made it understandable and easy to use. He is also credited with inventing the modern method of long **division** (see entry in volume 1).

Early career at St. John's College

Little is known about the details of Henry Briggs's family and early life. He received the typical schooling as a youngster, meaning that he learned Greek and Latin, and is said to have excelled in those subjects. Historians know that he had a brother, Richard, who became headmaster (principal) of a school in Norfolk, and that Briggs had literary friends. In 1577, Briggs entered St. John's College in Cambridge, and received his bachelor of arts degree from that school in 1581. Four years later, he was awarded his master of arts degree. He remained at St. John's and in 1592 became a lecturer of medicine and mathematics.

Becomes first professor of geometry at Gresham College

Henry Briggs

Briggs's training at St. John's College had obviously prepared him for both mathematics and medicine, but in 1596 he decided to focus solely on mathematics. That year, he accepted a position at the newly founded Gresham College in London and became that college's first professor of **geometry** (see entry in volume 1). He would remain in that position for the next twenty-three years. Besides teaching geometry, Briggs was interested in many of the practical aspects of his field, and became involved in using mathematics to help with navigation at sea. This naturally led him to the study of astronomy, and he became especially interested in the study of eclipses. The study of eclipses and astronomy in general required a great many long and involved calculations, and Briggs, probably like every other early seventeenth century astronomer, no doubt longed for a simpler and quicker way to do all the figuring that was required. Briggs worked on this problem and by 1610 he had published two books that contained tables that helped somewhat with these long and exhausting calculations. It is not surprising, therefore, that when Briggs first read about the new mathematical technique, which he described as "the noble invention of logarithms," according to *The MacTutor History of Mathematics Archive*, he immediately realized just how helpful logarithms could be.

Discovers Napier's invention of logarithms

Briggs first learned of logarithms in 1614 by reading *Mirifici logarithmorum canonis descriptio* (A Description of the Wonderful Canon of Logarithms), a book written by the Scottish inventor of logarithms, **John Napier** (1550–1617; see entry in volume 2). This work not only explained what logarithms were and how they worked, but also contained the first set of logarithmic tables. Briggs loved and marvelled at the beautiful simplicity of logarithms, which simplified complicated calculations by substituting the multiplication of large numbers with the simple addition of smaller ones. This became possible when Napier realized that large numbers could be more easily expressed in terms of "powers." For example, $100 = 10 \times 10$, which can be written as 10^2 or "ten to the power of two." The smaller number (2) above and to the right of the base number (10) is called an **exponent** (see entry in volume 1). With logarithms, as long as the same base number is used, one can multiply by *adding* exponents and divide by *subtracting* exponents. Briggs was delighted and amazed at Napier's achievement, saying, according to *The MacTutor History of Mathematics Archive,* that he "never saw a book which pleased me better or made me more wonder." He also could not understand why he himself had not thought of such a thing earlier.

Suggests improvements to logarithms and meets Napier

No sooner had Briggs read and understood Napier's ideas than he began teaching them at Gresham College. Being a bright and talented person, he also began trying to improve and advance Napier's methods. It was not long before Briggs arrived at what he thought was a considerable improvement, and wanted very badly to discuss it with the inventor himself. He therefore wrote to Napier, told him of his ideas, and received an invitation to visit. After finishing his lecture work at the end of the summer of 1616, Briggs made the long and difficult four-day trip from London to Edinburgh, Scotland, and arrived just as Napier was giving up hope that Briggs would ever come. Each mathematician was so touched and impressed by the other's presence that neither said a word for the first fifteen minutes. When he was able to speak, Briggs asked the great Napier how he had come to discover the idea of logarithms and stated that it was a wonder that no one else had done so before him. This was surprising, he said, since once logarithms are understood, one realizes that the idea itself is an amazingly simple and easy one.

During his stay with Napier, Briggs discussed his idea of making logarithms even easier to use. Briggs suggested that they create a table of logarithms using the base of 10, and Napier said he had already considered it. The two worked together on the idea while Briggs stayed with Napier for a month, and finally both agreed that having the base 10 was the best method. This system eventually became known as common or Briggsian logarithms. The first set of tables using common logarithms was published by Briggs in 1624 in his *Arithmetica logarithmica* (The Arithmetic of Logarithms), seven years after Napier's death. This volume provided the logarithms of the natural numbers from 1 to 20,000 as well as those for 90,000 to 100,000 computed to fourteen decimal places. Briggs's tables and the volumes later published by him and others proved to be so useful for performing the very large calcula-

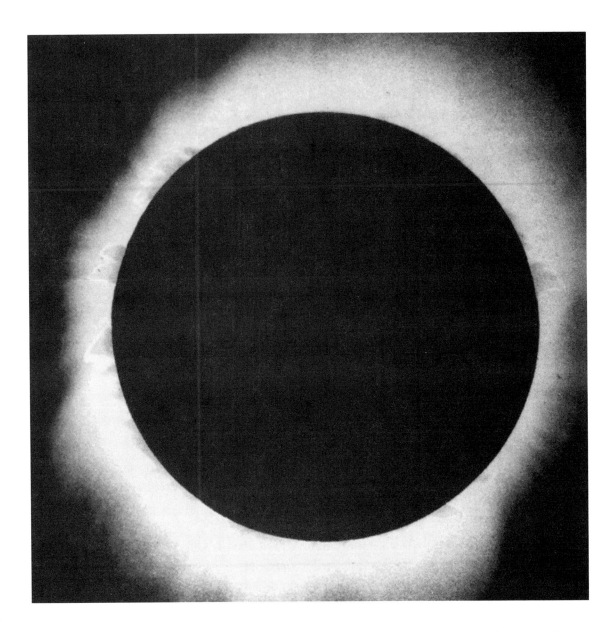

A view of the sun during an eclipse. Henry Briggs's use of logarithms helped in the study of eclipses. *Public domain.*

tions needed in astronomy and navigation that they were almost immediately accepted by everyone.

Moves to Oxford

Briggs became very well known and respected for his work with Napier. In 1620, English mathematician Henry Savile (1549–1622), the tutor of Queen Elizabeth I (1533–1603), created professorships of geometry and astronomy at Merton College in

Oxford, and he invited Briggs to become professor of geometry. Briggs accepted and left Gresham for Oxford that same year. While at Oxford, he continued to publish and teach until he died in 1630. Briggs was well liked by nearly everyone who knew him, and was described by his contemporaries as a humble, fair, and friendly person. It is also known that, unlike his colleague Napier, he considered astrology (predicting the future by studying the stars) to be a waste of time. As noted in *The MacTutor History of Mathematics Archive,* at his death, Briggs was described by English arithmetician **William Oughtred** (1574–1660) as "the mirror of the age for excellent skill in geometry."

For More Information

Abbott, David. *Mathematicians* New York: Peter Bedrick Books, 1985.

Burton, David M. *The History of Mathematics: An Introduction.* New York: McGraw-Hill, 1999.

Huxley, G. "Henry Briggs." *Biographical Dictionary of Mathematicians.* New York: Charles Scribner's Sons, 1991.

O'Connor, J. J., and E. F. Robertson. "Henry Briggs." *The MacTutor History of Mathematics Archive.* School of Mathematics and Statistics, University of St. Andrews, Scotland. http://www.groups.dcs.st-andrews.ac.uk/~history/Mathematicians/Briggs.html (accessed June 18, 2002).

Henry Briggs

Born 1620
Castle Lyons, Ireland

Died April 5, 1684
Westminster, London, England

English geometer

William Brouncker

William Brouncker.
Reproduced by permission of
Archive Photos, Inc.

William Brouncker is considered to be the first English mathematician to use continued fractions to express **pi** (see entry in volume 2). (Pi is the constant ratio of a circle's circumference to its diameter). He was a multitalented individual who served his king in many ways, and was one of the founders and first president of the Royal Society of London.

An aristocrat from the beginning

William Brouncker (pronounced BRUNG-ker) was part of the privileged English upper class. He was born in Castle Lyons, Ireland, to Sir William Brouncker and Lady Winefrid Leigh. His father held the position of Gentleman of the Privy (pronounced PRIH-vee; advisory board or council) Chamber to King Charles I (1600–1649), as well as a number of other important government positions. In September 1645, the king named Brouncker's father a viscount (pronounced VIGH-count; an English rank below an earl and above a baron). When Brouncker's father died two months later, that title passed to his oldest son, William, who automatically became the "Second Viscount Brouncker of Castle Lyons," and a high-ranking "peer of the realm" at the age of twenty-five. A peer is a member of one of England's five noble ranks: duke, marques, earl, viscount, and baron. Although little is recorded about his childhood and early education, Brouncker's parents probably followed the custom of the time and hired private tutors for William and his younger brother, Henry. It is known that Brouncker began his formal studies at Oxford University in London when he was sixteen years old. There, he showed a high ability for mathematics, languages, and medicine as well as a fondness for music.

English civil war

The year 1647 was significant for Brouncker: he received a Doctor of Physick (medical) degree and was forced into seclusion by dramatic changes that were taking place in England. His country had been experiencing a civil war between forces loyal to the king and those who wanted to overthrow the king and give power to the people's representatives known as Parliament (pronounced PAR-luh-munt; the English equivalent of the U.S. Congress). In 1646, the king's forces had lost the struggle, and he was put on trial. By 1649, King Charles I was executed and Oliver Cromwell (1599–1658), the opposition leader, had taken control of the country.

Cromwell and his followers rejected what was known as the "divine right" of kings, and his strong feelings against the upper class made this a tense and sometimes dangerous time for people of nobility, such as Brouncker. However, despite the fact that Brouncker was still loyal to the king, he apparently kept a low profile during these years and remained out of the public eye. He achieved this by focusing totally on his private studies of mathematics. Consequently, there is little recorded of Brouncker's life or activities between 1647 and 1660. The only thing historians know of during this time is the publication of Brouncker's *Musicae compendium* (Compendium of Music) in 1653. This was his only published work and was a translation of a book originally written by French algebraist, geometer, and philosopher **René Descartes** (1596–1650; see entry in volume 1).

Reemergence into public life

Brouncker reentered public life in 1660, when the king was restored to the throne. Charles I's son, Charles II (1630–1685), was now the king. With life in England now returning to the way

William Brouncker

it had been, Brouncker felt comfortable enough to enter Parliament. Then, as if to make up for lost time, he went on to hold several prominent offices: member of Parliament for Westbury in 1660; president of Gresham College from 1664 to 1667; commissioner for the navy from 1664 to 1668; comptroller of the treasurer's account from 1668 to 1669; and master of St. Catherine's Hospital in London from 1681 to 1684.

Mathematical work

Many people thought that Brouncker could have been one of the best mathematicians of his period if he had applied himself more fully to his mathematics. However, his fame as a mathematician rests mostly on his ability to solve problems that had been posed by others. Much of this work was actually done in correspondence with English mathematician John Wallis (1616–1703) who published some of Brouncker's work in his own books. For example, Brouncker worked on continued **fractions** (see entry in volume 1) and the expression for pi in response to a request from Wallis to come up with some type of expression of pi other than an infinite product. A product is the result of the **multiplication** (see entry in volume 2) of two factors or numbers. Brouncker was the first in England to use continued fractions, which have an infinite quotient. A quotient is the result when one number is divided by another. He used this technique to arrive at an approximation of the value of pi.

Royal Society of London

In 1662, King Charles II nominated Brouncker to become the first president of the Royal Society of London for the Improvement of Natural Knowledge. This is the oldest scientific society in England and is one of the oldest in Europe. His appointment by the king is significant, for at this time in England there were many very talented and highly respected scientists, any of whom was capable of being named the new society's first president. The king reappointed Brouncker every year until he resigned the presidency in 1677. Brouncker was known for his enthusiastic support of the notion that the society's members should do as much experimentation as possible (as opposed to doing more theoretical work or science for its own sake).

Brouncker was known very well by English writer Samuel Pepys (1633–1703), whose published diaries give historians an under-

William Brouncker

English writer Samuel Pepys, whose popular diaries included information about William Brouncker.
Courtesy of the Library of Congress.

standing of late seventeenth century England. Although some have written that Pepys was a close associate of Brouncker's and that he valued his friendship highly, Pepys also once wrote of how Brouncker had badly treated a lady friend of Pepys's saying, "I perceive he is a rotten-hearted, false man as any I know … and, therefore, I must beware of him accordingly, and I hope I shall." It was comments like this that made Pepys's diaries so popular when they were published. Whatever his true character, Brouncker never married, and when he died in 1684, his brother Henry inherited the title of viscount. Since Henry was also a bachelor, the title died with him in 1687.

William Brouncker

For More Information

Dubbey, John. "William Brouncker." In *Biographical Dictionary of Mathematicians.* New York: Charles Scribner's Sons, 1991, pp. 377–78.

O'Connor, J. J., and E. F. Robertson. "William Brouncker." *The MacTutor History of Mathematics Archive.* School of Mathematics and Statistics, University of St. Andrews, Scotland. http://www.groups.dcs.st-andrews.ac.uk/~history/Mathematicians/Brouncker.html (accessed June 21, 2002).

Westfall, Richard S. "William Brouncker." *Galileo Project.* http://es.rice.edu/ES/humsoc/Galileo/Catalog/Files/brounckr.html (accessed June 21, 2002).

Born February 28, 1552
Lichtensteig, Switzerland

Died January 31, 1632
Kassel, Germany

German algebraist, instrument maker, and astronomer

Joost Bürgi

Joost Bürgi is best known for developing his own version of algebraic logarithm tables. As a poorly educated but highly skilled instrument maker and astronomer, he independently invented a type of **logarithm** (see entry in volume 2) table about the same time as Scottish mathematician **John Napier** (1550–1617; see entry in volume 2). Since Bürgi published his invention several years after Napier, his discovery had little impact or influence.

Joost Bürgi.
Reproduced by permission of Mathematisches Forschungsinstitut Oberwolfach.

Early fame as an instrument maker

Little is known of the early years of Joost Bürgi (pronounced YOST BUEHR-ghee) since few records exist and none of his contemporaries wrote about his early life. Historians do know that he was born in Lichtensteig, Switzerland, which is east of Zurich and in the northern part of that small country. Most agree that since it is known Bürgi was unable to read Latin, it is reasonable to assume that he did not receive any sort of higher education. This is because at that time Latin was the universal language of education, especially scientific education. However, since it is also known that Bürgi was appointed court watchmaker to a duke when he was twenty-seven years old, historians can assume that he must have learned the trade of instrument-making by working for a master toolmaker as an apprentice (someone who serves under a skilled worker and is taught his trade). In the middle of the sixteenth century, instrument making was a high skill and was not something everyone could do. Also by then, the growth of trade, the early beginnings of industrialization or manufacturing, and the rise of science and its needs for better tools and more accurate measuring devices combined to make someone with Bürgi's skills be in high demand. Good instrument makers were therefore never out of work, and the best were usually selected by royalty and the ruling classes for whom they practiced their craft.

It was probably in this manner that Bürgi was made the court watchmaker to Duke Wilhelm IV (1532–1592). This appointment was surely a high honor for Bürgi. Many think that working in the duke's royal observatory also gave Bürgi the opportunity to complete his education or at least to study and learn more. From 1579 to 1592, he worked at the royal observatory making clocks and other precision instruments to aid astronomical observation. While there, he also invented or improved devices for use in practical **geometry** (see entry in volume 1), which is a geometry done for a functional purpose. One of these devices, called the military geometric proportional compass, served the same purpose as the modern slide rule. His version of this device, which was used to make measurements and calculations, was said to be as good as that designed and built by Italian mathematician, astronomer, and physicist **Galileo Galilei** (1564–1642; see entry in volume 1), who lived at the same time.

Joins the emperor's court

By the end of the sixteenth century, Bürgi's reputation for instrument making attracted the attention of Rudolf II (1552–1612), emperor of the Holy Roman Empire. "Holy Roman Emperor" is a title used by modern historians to describe the king who ruled Germany and the territory of some other countries around it (parts of what are now Austria; the Czech Republic; Switzerland; France; the Low Countries of the Netherlands, Belgium, and Luxembourg; and even some of Italy). Rudolf II was fascinated by science and wanted to establish a science center in Prague (pronounced PRAHG), where his court was located. So he set out to try to attract the best

and most famous scientists in Europe. After Duke Wilhelm IV died, Bürgi accepted Rudolf's invitation—which when it comes from an emperor can also be considered a command—and joined the other scientists at the Prague center in 1603.

When Bürgi moved to Prague to take up the position of imperial clockmaker, he found himself in the company of several very learned men who also worked for the emperor. One of these was German astronomer and mathematician **Johannes Kepler** (1571–1630; see entry in volume 2). Bürgi worked with Kepler as an assistant, refining instruments and performing mathematical calculations. As he had done at Wilhelm's court, he took advantage of the resources Rudolf had assembled at his observatory to continue his own education. Kepler had the title of imperial mathematician and employed Bürgi primarily as his "computer," meaning that Bürgi was in charge of performing the many long and detailed calculations Kepler required. Bürgi was especially qualified to do this since during his stay under Wilhelm, his interest in astronomical calculations

German astronomer Johannes Kepler (left) talks with Rudolf, emperor of the Holy Roman Empire. *Reproduced by permission of the Corbis Corporation.*

inspired him to try to develop faster and more accurate ways to compute. Before Bürgi joined Rudolf's court, he had already achieved a great deal in this regard.

Joost Bürgi

Independently invents logarithms

Bürgi was not someone who philosophised or theorized about mathematics. Instead, he was a very practical man who focused on making the application of mathematics to astronomy faster and more reliable. At Wilhelm's court, he recognized the need for an easier method of multiplying large numbers, and in 1584 began work on improving a system of computing called "prosthaphaire-sis" (pronounced prahs-thuh-fair-EE-sis). This was the name that astronomers used to describe a hoped-for method of doing compli-cated **multiplication** (see entry in volume 2) by simple **addition** (see entry in volume 1). Sometime by the end of the 1580s, Bürgi happened upon the idea of logarithms, and he created some actual conversion tables by the time he went to Prague.

Logarithms are numbers known as **exponents** (see entry in volume 1) that are used to express repeated multiplications of a single num-ber. A logarithm is always used in reference to a number or "power" to which another "base" number is raised. In simple terms, loga-rithms reduce the multiplication of large numbers to the simple addition of small numbers. The key is found is realizing that large numbers can be expressed more easily in terms of "powers," so that 100, which equals 10 times 10, can be written as 10^2 or "ten to the power of two." As long as the base number is the same, logarithms can be multiplied by *adding* exponents or powers and divided by *subtracting* exponents. Bürgi brought some version of this comput-ing shorthand with him when he went to Prague in 1603, for schol-ars know that he already had it written down in a manuscript by then. The practical Bürgi apparently had no interest in establishing himself as a great mathematician, so he never looked to publish his invention or to make it better known. However, when Bürgi moved to Prague and Kepler learned of Bürgi's breakthrough, Kepler even-tually persuaded Bürgi to prepare his work for publication. His log-arithm tables were not printed until 1620, and by that time, Scot-tish mathematician John Napier had already published his own independent discovery of logarithms in 1614. Although Napier's method was somewhat different from Bürgi's, both were differing versions of the same basic idea. Since Napier was first in publishing, Bürgi's work had no impact or influence.

As a talented and hard working individual, Bürgi is credited with several other accomplishments. He is known to have used the new decimal point, which was a major breakthrough since it was used to separate a **whole number** (see entry in volume 2) from the fraction part of a number. In addition to the military geometric proportional compass, Bürgi also invented a double compass and a special instrument for perspective drawing, which allowed artists to give their paintings a realistic third dimension. Many also credit Bürgi with building the first clock with a separate minute hand around 1577. Finally, historians know that Bürgi was married twice and that German mathematician Benjamin Bramer (1580–1650) was his foster son. Bürgi ended his career in Kassel, Germany, to which he returned after leaving the court for good in 1631, a year before he died.

For More Information

Boyer, C. B. *A History of Mathematics.* 2d ed. Revised by Uta C. Merzbach. Princeton, NJ: Princeton University Press, 1985, pp. 309–11, 327–328, 345–50.

Novy, Lubos. "Joost Bürgi." In *Biographical Dictionary of Mathematicians.* New York: Charles Scribner's Sons, 1991, pp. 389–90.

O'Connor, J. J., and E. F. Robertson. "Joost Bürgi." *The MacTutor History of Mathematics Archive.* School of Mathematics and Statistics, University of St. Andrews, Scotland. http://www.groups.dcs.st-andrews.ac.uk/~history/Mathematicians/Burgi.html (accessed June 21, 2002).

Joost Bürgi

Born August 16, 1821
Richmond, Surrey, England

Died January 26, 1895
Cambridge, England

English algebraist and geometer

Arthur Cayley

Arthur Cayley.
Courtesy of the Library of Congress.

A rthur Cayley was one of the most productive mathematicians in the history of mathematics. A brilliant mathematician who sincerely loved his field, Cayley single-handedly raised English mathematics to a higher level. He also did advanced work in **geometry** (see entry in volume 1) that helped pave the way for such twentieth-century breakthroughs as the theory of relativity and quantum mechanics.

Brilliant from the start

Arthur Cayley (pronounced KAY-lee) was born in England, but he could just as easily have been born in Russia. His parents, Henry Cayley and Maria Antonia Doughty, were English but they lived in St. Petersburg, Russia, where Cayley's father was a successful merchant. Cayley was born during a short family visit to England, but he and his family returned to Russia within a few months of his birth and he would spend the next eight years of his life there. By the time his parents left Russia for good and returned to England, it was very clear that their son was no ordinary student. His ability in mathematics was astounding, and they believed that he was already a genius. Unlike most children, he spent his playtime happily solving mathematical problems.

As a youngster, Cayley was sent to a small private school where he stayed until he was fourteen years old. By then, he was more than ready for higher learning, so he was enrolled in King's College School in London. There, his mathematical ability really shined through and his teachers knew that he needed more than they could give. They suggested to his father that the teenaged Cayley not be made to enter the family business as a merchant but instead be allowed to pursue his natural mathematical ability as far as it would take him. At seventeen years old, Cayley entered Trinity College in Cambridge, where he was immediately recognized as a mathematical genius. As an undergraduate, he published three papers and graduated at the top of his class in 1842, having placed first on the university's final examinations in mathematics. In fact, he did so well that his professors allowed Cayley to skip the oral part of his exam since it was obvious that he knew more than they did.

Switches to law

After graduating, Cayley remained at Cambridge where he taught mathematics for three years. However, Cayley had taken an oath to study divinity or religion when he was admitted to Cambridge, and he realized that he would soon be required to become a clergyman in the Anglican Church. Since he was unwilling to do this and did not want to be compelled to focus on his religious education, Cayley decided to give up his teaching position. This meant, however, that he would have to find another profession. He decided to study law and entered the historic Lincoln's Inn in London. Researchers know that Cayley had a phenomenal memory and could remember anything he read if he wished to do so, and this probably helped him succeed in law. In 1849, he was admitted to the bar. He practiced law for the next fourteen years, and he made quite a good living at it. However, Cayley was the first to admit that he considered the law simply to be a means of making money so that he could pursue what he always loved—mathematics.

However Cayley did not just "pursue" mathematics in his spare time. In fact, he gladly spent every free moment on mathematics and somehow found the time to write nearly three hundred papers in the fourteen years he practiced law. This is a phenomenal amount for someone who could not devote all his time to mathematics, and some of the work he did during those years is considered his most original.

Arthur Cayley

Arthur Cayley

In 1863, Cayley married Susan Moline (with whom he would have two children) and was offered the Sadlerian Professorship of Pure Mathematics at Cambridge. This was a highly honored position but it did not pay very well. As Cayley had just married, he could have chosen to remain in the law, but instead, he gave up the large salary he had earned as a lawyer in order to work in the field he loved. For the rest of his life, Cayley remained at Cambridge in the field of mathematics.

Goes on to incredible productivity

Now firmly entrenched in mathematics, Cayley went to work. During his lifetime, he published nearly one thousand papers and studied and wrote about almost every aspect of mathematics. Today, his collected works total thirteen volumes. Cayley's favorite area of mathematics was pure mathematics since he did not necessarily try to solve a problem that had a practical, everyday application. Instead, his mind roamed far beyond the mathematics of his day, with the result that he founded or developed some of the more specialized (and least understood) branches of mathematics. Cayley is now recognized as the developer of a theory of matrices (pronounced MAY-trih-sees). A matrix (the singular version of matrices) is any rectangular array of numbers, meaning that it has rows (horizontal) and columns (vertical). Although this useful device was known and used long before Cayley, he was the first to arrive at an abstract definition of a matrix and to create a theory of how they work and how they should be used. As a branch of higher mathematics, matrix theory is used to solve many real-world problems that require the simultaneous solution of five, ten, twenty, or even one hundred unknowns. Matrices have become standard tools in fields as different as education, psychology, chemistry, physics, statistics, economics, and even electrical and aeronautical engineering.

Cayley was also the first to discover what is called "n-dimensional geometry," which German American physicist and mathematician **Albert Einstein** (1879–1955; see entry in volume 1) eventually used to study space-time and develop his theory of relativity. Cayley's work on matrices would even later serve as the foundation for the quantum mechanics of German physicist Werner Karl Heisenberg (1901–1976). The work of Einstein and Heisenberg was essential for the twentieth century revolution in modern physics.

Honored and respected in his lifetime

Cayley was not just an extremely prolific mathematician, but was also a highly original thinker who made a staggering number of contributions to mathematics. He was known for his ability to make complicated ideas understandable, and his writing style was always direct, methodical, and clear. During his long life, Cayley received nearly every major award of his time, including the De Morgan medal given by the London Mathematical Society and the Copley medal of the Royal Society.

As an individual, Cayley was interested in fields other than mathematics. For example, he read thousands of novels, not only in English but also in French, Greek, German, and Italian. He was a noted speaker and educator as well as a very good painter of watercolors. He also took a great interest in architecture and architectural drawing. As a student of botany (the study of plants), he enjoyed nature and was also an accomplished mountain climber. He once said that he enjoyed mountain climbing, despite how difficult and tiring it was, because when he reached the top, the feeling he had was similar to what he experienced when he solved a difficult mathematical problem. Given all the work he had to do to solve an intricate mathematical problem, he said that climbing mountains was actually easier.

Cayley was described as a courteous and gentle man, and one who had a frequent smile throughout his life. Overall, Cayley was a true Renaissance man (a person of very broad interests and accomplishments), but to the people of his profession, he was called "the mathematician's mathematician."

For More Information

"Arthur Cayley." http://www.stetson.edu/~efriedma/periodictable/html/C.html (accessed June 21, 2002).

Mathematicians and Computer Wizards. Detroit: Macmillan Reference USA, 2001.

North, J. D. "Arthur Cayley." In *Biographical Dictionary of Mathematicians.* New York: Charles Scribner's Sons, 1991, pp. 450–58.

O'Connor, J. J., and E. F. Robertson. "Arthur Cayley." *The MacTutor History of Mathematics Archive.* School of Mathematics

Arthur Cayley

Arthur Cayley

and Statistics, University of St. Andrews, Scotland. http://www.groups.dcs.st-andrews.ac.uk/~history/Mathematicians/Cayley.html (accessed June 21, 2002).

Born March 24, 1948
Ji'an (Chi-an), China

Chinese-born American analyst

Sun-Yung Alice Chang

S un-Yung Alice Chang's work with a highly specialized type of geometry (see entry in volume 1) has been recognized by her peers as having made "deep contributions" to her field. Born in China during the Communist revolution, she overcame having to travel to a new land and learn a new language and became a high achiever who has received many major mathematical awards and honors.

From China to America

Sun-Yung Alice Chang was born in Ji'an, China, which is in the southern part of that huge country and a couple of hundred miles from its east coast. She was born in 1948, a year in which China was experiencing a revolution. Following World War II (1939–45), Communist forces led by Mao Tse-tung (pronounced MOWD-ZUH-DOUNG; 1893–1976) challenged China's Nationalist army. By 1949, the Communist forces had won the civil war, and the Nationalist leaders fled to an island off the southeast coast of China called Formosa (now called Taiwan; pronounced tie-WAHN). There, the Nationalists set up a separate government from China in order to continue the fight. Chang's father, Fann Chang, and her mother, Li-Ching Chern, took their

Sun-Yung Alice Chang

daughter and moved to Taiwan after the Chinese revolution, and it was there that young Chang grew up. A good student, Chang was able to enter the National University of Taiwan and received her bachelor's degree from that institution in 1970. Very soon after, she decided to move to the United States, and she applied to the University of California, Berkeley, for graduate studies. She was accepted and earned her Ph.D. in mathematics in 1974.

Increasingly better positions

Upon receiving her Ph.D., Chang began accepting a string of increasingly prominent teaching positions throughout the United States. First, she served as assistant professor at the State University of New York–Buffalo for the academic year 1974–75. In 1975, she went to the west coast and accepted a position as the Hedrick Assistant Professor at the University of California, Los Angeles (UCLA). She remained there for two years and made another cross-country trip to teach at the University of Maryland, College Park. In 1980, she was back on the other coast again at UCLA where she became an associate professor. By 1982, she was named a full professor there.

In 1986, Chang achieved national recognition when she delivered an address at the International Congress of Mathematicians. She spoke of her mathematical specialty, which could be described as a subset of differential geometry. Differential geometry applies calculus equations to points on a geometrical figure and creates a more general picture of that figure's geometrical characteristics. Since differential geometry deals with certain types of curved surfaces, it was originally used to survey portions of the Earth's surface. Today, however, it has proven highly useful in mathematical physics, especially for **Albert Einstein**'s (1879–1955; see entry in volume 1) theory of relativity with its notion of "curved" space-time.

While at UCLA, she was very active in the mathematical community and received numerous awards. In 1988, she was honored as the UCLA Outstanding Woman of Science. During those years, she served as editor of several mathematical journals, and was vice president of the American Mathematical Society for two years. In 1995, Chang received the prominent Ruth Lyttle Satter Prize, which she was given at the 101st annual meeting of the American Mathematical Society. Chang was honored for her major contributions to her highly specialized field, and for her commitment to encourage women in science. In her acceptance speech, Chang acknowledged her debt to her collaborators and promised to continue her mathematical work in her obscure field. Chang was deeply honored by this award, and her speech referred to the role of women in mathematics and her experiences in particular. She stated that it was very important for her to have had so many good female role models when she was in school. She recalled that in her undergraduate class in Taiwan, there were twelve women in a class of forty mathematics majors. As noted in *The MacTutor History of Mathematics Archive,* Chang said, "I can personally testify to the importance of having role models and the companionship of other women colleagues." She believes that women and men are equally talented in mathematics, and she sees a better future for women in the math field as more and more women are becoming active research mathematicians.

Chang remained at UCLA until 1998, when she took a professorship at Princeton University in New Jersey. She has remained at Princeton and continues to work on new and different techniques in her specialized field. She also has dedicated herself to improving the status of women in her profession. Further, she always finds the time to involve her students in some way in her field, and has recently taken part in a lecture program that targets an audience of young graduate students. She continues to be active with the American Mathematical Society, working on a range of committees, including its editorial board, and speaking at a number of its meetings.

In 1973, Chang married Paul Yang, who is also a mathematician and a scientific collaborator. The couple has a daughter and a son. In her spare time, Chang enjoys reading novels, taking walks, and listening to music.

For More Information

Mathematicians and Computer Wizards. Detroit: Macmillan Reference USA, 2001.

O'Connor, J. J., and E. F. Robertson. "Sun-Yung Alice Chang." *The MacTutor History of Mathematics Archive.* School of Mathematics and Statistics, University of St. Andrews, Scotland. http://www.groups.dcs.st-andrews.ac.uk/~history/Mathematicians/Chang.html (accessed June 24, 2002).

Sun-Yung Alice Chang

Sun-Yung Alice Chang

"Sun-Yung Alice Chang." *Profiles of Women in Mathematics.* http://www.awm-math.org/noetherbrochure/Chang01.html (accessed June 24, 2002).

Born December 5, 1895
Evansville, Indiana

Died November 28, 1969
Washington, D.C.

African American pure mathematician and educator

Elbert F. Cox

Elbert F. Cox was the first African American and possibly the first person of African heritage in the world to earn a Ph.D. in mathematics. During his career, he rose from high school instructor to chairman of Howard University's department of mathematics. A scholarship fund was named after him, and the National Association of Mathematicians named and started its annual Cox-Talbot Address in his honor.

Role models at home

Elbert Frank Cox was born on December 5, 1895, in a racially mixed neighborhood in Evansville, Indiana. He was the oldest of three boys born to Johnson D. Cox and Eugenia D. Talbot. Johnson Cox was an elementary school principal who graduated from Evansville College and had done graduate work at Indiana University. The Cox family was close, very religious, and had a deep respect for learning that reflected the father's educational career. When Elbert demonstrated in high school that he had unusual ability both in mathematics and physics, his father directed Elbert toward Indiana University, where he had conducted some of his own graduate studies. At Indiana, Cox majored in mathematics,

Elbert F. Cox

was elected to a student office, and joined the Kappa Alpha Psi fraternity. When he was awarded his bachelor's degree from Indiana in 1917, the United States was already involved in World War I (1914–18), so after graduation he entered the U.S. Army as a private. Within six months, he was promoted to staff sergeant, and served overseas in France. After the war, Cox returned home to pursue a teaching career and became an instructor of mathematics at a high school in Henderson, Kentucky.

Pursues pure mathematics

Soon after he started teaching at a high school, Cox joined the faculty of Shaw University in Raleigh, North Carolina, where he taught for two years before he applied for admission to Cornell University in Ithaca, New York, in December 1921. It was his outstanding performance at Shaw University that enabled him to apply for and earn the Erastus Brooks Fellowship, which would lead to a Ph.D. at Cornell. At the time, Cornell was one of only seven universities in the United States that offered a doctoral program in mathematics. Cox received his full scholarship in September 1922 and began his graduate studies at Cornell. According to *Mathematicians of the African Diaspora,* during the process of application, one of the persons who served as a reference wrote a very positive, supportive letter for Cox, but also said he anticipated "certain difficulties for the young man because of the fact that he is of the colored race." There was no mistaking Cox's race, however, as even the transcript he received after graduating from Indiana University had the word "colored" (the term for African Americans at the time) in large letters printed across it.

A long, productive career in teaching

Whatever "certain difficulties" Cox may have encountered at Cornell, they were not enough to prevent or even slow down his progress toward his doctorate. In the summer of 1925, when Cox received his Ph.D. in pure mathematics from Cornell, he made history as the first of his race to earn a Ph.D. in mathematics. This field is considered by some to be especially difficult since it is concerned with mathematical theory rather than the practice or application of mathematics. His accomplishment was an achievement for Cornell as well, since it had awarded its very first Ph.D. in mathematics only a generation before in 1886. Cox was also fortunate to have Lloyd Williams (1888–1976), founder of the Canadian Mathematical Society, as his thesis advisor at Cornell. When Williams realized that if Cox earned his Ph.D. in mathematics he would become the first African American to do so, he also recognized that his young student would also probably be the first of his race in the world to do so. This led Williams to encourage Cox to send his doctoral dissertation to a university in another country so that Cox's claim would be recognized internationally. Although Cox's dissertation was not accepted in England and Germany, it was accepted by Japan's Imperial University at Sendai, now called Tohoku University.

In September 1925, Cox took his first postgraduate position and became head of the mathematics and physics department at West Virginia State College. Four years later, he moved to Washington, D.C., to join the faculty of Howard University. Cox would remain at Howard for the remainder of his career, becoming the chairman of its department of mathematics in 1947. He held this position until 1961, when a university rule forced him to resign at the age of sixty-five. He remained at Howard as a full professor for another five years, and retired in 1966. During Cox's full career at Howard, he enjoyed the company of several major African American mathematicians, including Dudley Woodard (1881–1965) and William Claytor (1908–1967), the second and third African American Ph.D.s. Cox also had considerable influence in making Howard University become recognized as black America's principle place of learning.

America in 1925

In the twenty-first century, it is often difficult to imagine what America was like for an African American in the 1920s. Certainly slavery was long gone from this country, but African Americans still lived under the constant cloud of racism, which worked to always remind them that they were second-class citizens. In 1925, when Cox received his Ph.D. in mathematics, only twenty-seven other Ph.D.s were awarded nationwide in mathematics. Incredibly, however, that number was exceeded by the number of African American men who were lynched that year—thirty-one. For Cox to survive and even excel in such a climate stands as a wonderful testament to both him and the people and institutions who helped him achieve his goal.

Elbert F. Cox

Elbert F. Cox

Cox also had time for a family. He married Beulah P. Kaufman, an elementary school teacher, on September 14, 1927, and they had three sons, James, Eugene, and Elbert. Cox died after a brief illness in 1969. To honor Cox, Howard University's Mathematics Common Room hung a portrait of him as a reminder of his personal achievements and of the contributions he made to that university.

For More Information

Harris, Lissa. "Math Department Honors CU Pioneer Elbert Cox, First Black Math Ph.D." *Cornell Chronicle.* http://www.news.cornell.edu/Chronicle/02/2.28.02/ElbertCox.html (accessed June 24, 2002).

Houston, Johnny L. "Elbert Frank Cox." *MAA Online.* http://www.maa.org/summa/archive/Cox_EF.htm (accessed June 24, 2002).

Williams, Scott W. "Elbert F. Cox." *Mathematicians of the African Diaspora.* http://www.math.buffalo.edu/mad/PEEPS/cox_elbertf.html (accessed June 24, 2002).

Born October 6, 1831
Braunschweig (now Brunswick), Germany

Died February 12, 1916
Braunschweig, Germany

German number theorist

Richard Dedekind

Richard Dedekind was an original mathematical thinker who was in many ways ahead of his time. He gave a new understanding to the concept of **irrational numbers** (see entry on Rational and Irrational Numbers in volume 2) and was a pioneer of the abstract **algebra** (see entry in volume 1) of the twentieth century. He also contributed notably to mathematics by editing the collected works of noted mathematicians Peter Gustav Lejeune Dirichlet (1805–1859), **Carl Friedrich Gauss** (1777–1855; see entry in volume 1), and **Bernhard Riemann** (1826–1866; see entry in volume 2).

A distinguished family and a solid education

Julius Wilhelm Richard Dedekind (pronounced DAY-duh-kihnt) was born in what is now Brunswick, Germany. He was the youngest of four children born to Julius Levin Ulrich Dedekind, a lawyer and a professor of law, and Caroline Marie Henriette Emperius, the daughter of a professor. As an adult, Dedekind dropped his first two names and went simply by Richard. Starting school when he was seven years old, Dedekind attended the Gymnasium (pronounced gihm-NAH-zee-um; a nine-year

Richard Dedekind

school run by the state that prepares pupils for higher education) in Brunswick. Dedekind studied there until he was sixteen and concentrated mainly on physics and chemistry, thinking of mathematics only as a handy scientific tool. When he entered Collegium Carolinum (Caroline College) in 1848 (a pre-university institution), he became more interested in mathematics and spent his two years there obtaining a solid foundation in mathematics. In 1850, he enrolled at the University of Göttingen and arrived much better prepared than most students.

Although prominent German mathematician Carl Friedrich Gauss was still a professor at Göttingen, the university was not yet as influential as it would become to mathematics. However, Gauss was a major figure in the world of mathematics, and Dedekind was fortunate enough to become the elderly man's last doctoral student. Fifty years after Dedekind took his first course with Gauss, Dedekind remembered the lectures as the most beautiful he had ever heard. As an indication of the kind of excellence he strove for, when Dedekind was awarded his Ph.D. in 1852 after only four semesters of work, he felt he was not yet ready to become a professional mathematician because he still was not as well trained and knowledgeable in advanced mathematics as he wanted.

Takes posts at Göttingen, Zürich, and Brunswick

In an effort to learn more about mathematics, the first thing Dedekind did after obtaining his doctorate was to become a student once again. Knowing that there was more for him to learn than what he had obtained at Göttingen, Dedekind spent the next two years filling the gaps in his education by attending the lectures of some of the best mathematical minds at the University of Berlin. By the summer of 1854, he finally felt qualified to teach and began as a lecturer at Göttingen. When Gauss died in 1855, German number theorist and analyst Peter Gustav Lejeune Dirichlet was appointed to fill the vacancy. This appointment would prove to be a landmark in Dedekind's life as he not only became inspired by Dirichlet but became his close personal friend as well. Amazingly, as noted in *The MacTutor History of Mathematics Archive,* Dedekind later wrote that it was his daily contact with Dirichlet that finally made him feel that, "I am for the first time beginning to learn properly."

German number theorist and analyst Peter Gustav Lejeune Dirichlet, associate and close friend of Richard Dedekind. *Reproduced by permission of Archive Photos, Inc.*

Richard Dedekind

Inspired by Dirichlet and finally feeling confidence in his knowledge of mathematics, Dedekind applied for the vacant chairman's position at the Polytechnic School in Zürich, Switzerland. He soon received the appointment and began teaching there in autumn 1858. This type of position in Zürich was traditionally a first step toward a professorship in one of the better universities in Germany, but after five years in Zürich, the opportunity arose for Dedekind to return to his beloved city of Brunswick. The old Collegium Carolinum had been upgraded from a technical school to a

Richard Dedekind

polytechnic (pronounced pah-lee-TEK-nik) school (a higher level of instruction in many technical arts and applied sciences). In 1862, Dedekind accepted a professorship in the same school where his father had worked, and settled into a career at the Collegium in which he would remain for the rest of his life.

Mathematical contributions

It was not until Dedekind went away to teach in Zürich that he began to work on original mathematics. In 1858, what eventually came to be called a "Dedekind cut" first occurred to him. This is a completely original idea that treats the problem of irrational numbers in an entirely new manner. An irrational number is expressed as a nonrepeating **decimal** (see entry in volume 1) **fraction** (see entry in volume 1) which, when carried out, simply goes on forever (in contrast to a rational number, which comes out even with no remainder or has a repeating decimal like .33333). It is because of Dedekind, therefore, that there is now a device to categorize irrational numbers as fractions.

On the basis of Dedekind's work with irrational numbers alone, he is considered among the greats of mathematics. However, Dedekind had a long and very productive professional life, and he contributed several other major mathematical concepts and ideas that were elaborated on by others. Besides his own important publications, he published some of the manuscripts of Gauss, and was one of the few people capable of commenting on them with real understanding. He also edited the work of his friend Dirichlet and German geometer and mathematical physicist Bernhard Riemann. Finally, although he was not a great lecturer, he was able to write about mathematics better than nearly anyone. In fact, his manner of explaining mathematics in writing was so clear and understandable that nearly every mathematician was able to grasp his ideas immediately. This method of writing not only became the model for everyone, but it actually inspired a new style of mathematics that later generations would seek to follow.

Personality and character

Dedekind is considered among the top mathematicians of all time, yet he was a man who never sought fame or even popularity. He spent nearly his entire life teaching in his native Brunswick, leaving only to take holidays in Switzerland, Austria, or the Black Forest in southern Germany. He never married and lived with his sister, Julia. His brother lived nearby as well. He seemed content to live his life on a small scale, and spent considerable time with his own relatives and the families of his close friends, especially Dirichlet. He was a remarkably healthy man who also displayed considerable musical talent. He played both the piano and the cello expertly, and he even composed an opera. Like many of his close friends and colleagues, he was a man of strong will who stood by what he believed in and did not compromise. Like them also, he chose to live a simple, orderly life. He was described as warm-hearted and humorous, but with a strictness and sense of duty that he always applied to himself.

A good example of his attitude and style is seen when a major publisher issued a calendar of famous mathematicians and listed his death date as September 4, 1899. Amused by this, the sixty-eight-year-old Dedekind—who would live another sixteen years—wrote to the editor that although the day of his death might be correct, he was sure the year was wrong. On that date and year, he said, he had enjoyed a delightful lunch with his good friend, German analyst and set theorist **Georg Cantor** (1845–1918; see entry in volume 1). Dedekind was a truly modest man who always pursued and loved truth, and once said of himself, "For what I have accomplished and what I have become, I have to thank my industry [willingness to work hard] … rather than any outstanding talent."

For More Information

Abbott, David. *Mathematicians.* New York: Peter Bedrick Books, 1986.

Biermann, Kurt R. "Julius Wilhelm Richard Dedekind." In *Biographical Dictionary of Mathematicians.* New York: Charles Scribner's Sons, 1991, pp. 576–81.

Franceschetti, Donald R., ed. *Biographical Encyclopedia of Mathematicians.* New York: Marshall Cavendish, 1999.

"Julius Wilhelm Richard Dedekind." http://www.stetson.edu/~efriedma/periodictable/html/Db.html (accessed June 24, 2002).

Richard Dedekind

Richard Dedekind

O'Connor, J. J., and E. F. Robertson. "Julius Wilhelm Richard Dedekind." *The MacTutor History of Mathematics Archive.* School of Mathematics and Statistics, University of St. Andrews, Scotland. http://www.groups.dcs.st-andrews.ac.uk/~history/Mathematicians/Dedekind.html (accessed June 24, 2002).

Schlager, Neil, ed. *Science and Its Times.* Vol. 5. Detroit: Gale Group, 2001.

Born April 9, 1919
Philadelphia, Pennsylvania

Died June 3, 1995
Bryn Mawr, Pennsylvania

American computer engineer

J. Presper Eckert

J. Presper Eckert was the co-inventor with John W. Mauchly (1907–1980) of the world's first fully electronic digital **computer** (see entry in volume 1) known as ENIAC. Together, they also produced the first commercial digital electronic computer, UNIVAC, and began a worldwide computer revolution that continues to change the world in profound ways.

J. Presper Eckert.
Reproduced by permission of AP/Wide World Photos.

A privileged and sophisticated boyhood

John Presper Eckert Jr. was born in Philadelphia, Pennsylvania, the only child of John Presper Eckert and Ethel Hallowell. His father was a self-made millionaire, having done extremely well as a real estate developer and contractor. Young Eckert grew up in the Germantown section of Philadelphia and attended the William Penn Charter School there, the oldest private boys' school in the United States. In addition to these privileges, by the time he was twelve years old, Eckert had traveled over 125,000 miles, visiting the forty-eight states, Alaska (not yet a state), and most of the great cities of Europe. He even made a trip to Egypt. Eckert was able to do this at such a young age by accompanying his father on business trips as well as going on frequent family vacations. Occasionally, he would even find himself in the company of movie stars such as Douglas Fairbanks Sr. (1883–1939) and Charlie Chaplin (1889–1977), as well as politicians such as President Warren G. Harding (1865–1923). This was no ordinary childhood, and Eckert proved to be no ordinary child.

At the Penn Charter School, Eckert was an excellent student. He was regarded as a mathematics whiz and an electronics genius. As a youngster, he had already built his own radio and was always tinkering with some sort of electronic gadget. Upon graduation from Penn Charter, Eckert had hoped to attend the Massachusetts Institute of Technology (MIT), but his mother had other ideas. Eckert most certainly would have been admitted to MIT since he had achieved the second highest score in the country on his college board mathematics test. However, since he was his mother's only child, she did not want him to attend school so far from home. To convince Eckert to apply to a school close to home, his father made up a story that the family could not afford the tuition at MIT. Amazingly, young Eckert believed this and decided to please his parents, settling finally on the Moore Engineering School at the University of Pennsylvania. When Eckert finally discovered the truth during his freshman year, he became very angry and had a bad year academically. He remained at the Moore School, however, and earned his undergraduate degree in electrical engineering in 1941.

Meets John Mauchly

After graduation, Eckert decided to remain at the university to teach and attend graduate school. It was while he was working as a lab instructor that he met thirty-four-year-old American physicist and engineer John W. Mauchly (pronounced MAWK-lee). Although twelve years older than Eckert and an established professor of physics, Mauchly was taking an eight-week government-paid defense course in electronics, and Eckert happened to be teaching it. After they met, the two men soon realized that they shared similar interests in what was then called computational devices or computers. Mauchly is said to have been the idea man, and Eckert was the one with know-how. It is not surprising that Eckert and Mauchly began talking, since Eckert was described by Robert Slater in his book *Portraits in Silicon* as being someone "who liked to work things out orally in the presence of someone; it didn't matter whether it was a technician or a night watchman. He was highly nervous and would rarely sit in a chair or stand still

What Led to Eckert-Mauchly Meeting

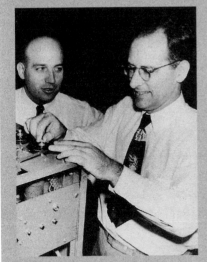

J. Presper Eckert (left) and
John W. Mauchly.
*Reproduced by permission of AP/Wide
World Photos.*

The Eckert-Mauchly team would never have happened had John Mauchly not attended the Moore School of Electrical Engineering at the University of Pennsylvania in 1941 where J. Presper Eckert was just starting to teach. However, there was no obvious or reasonable explanation why Mauchly was a student there since, at the time, he was the chairman of the department of physics at Ursinus College, a small educational institution located appropriately in Collegeville, Pennsylvania. The year 1941, however, was not typical, for as the Germans took over Europe and threatened England, Japan was already planning its attack on the United States's fleet at Pearl Harbor in Hawaii.

Since the American military saw war as inevitable, they knew they would soon need many young engineers who had been trained to operate the new electronic weapons and communications systems they were already developing. The Moore School was one institution that was doing this kind of training, and it sent Mauchly a letter asking whether he knew of any students who might be good candidates for such training. Since Mauchly at this point had already conceived the idea of an electronic version of a calculator, he knew that as a physicist he would need to learn more about electronics. So, even though Mauchly held a Ph.D. in physics and was the chairman of the department, he enrolled himself in the program. It was not long before Mauchly, the man of vision and ideas, met Eckert, the technical wizard who could make those dreams a reality. Together they would build the first electronic digital computer.

while he was thinking. Often he would crouch on top of a desk or pace back and forth."

By 1943, when Eckert received his master's degree, the two men were ready to join forces and work toward their goal of designing and building an electronic computer. Both were well aware that American mathematician **Howard H. Aiken** (1900–1973) had produced the first large-scale automatic digital calculator called Mark I in January 1943. However, both also knew that it was mostly electromechanical (meaning that it used magnetically-operated devices, since it had switches, clutches, shafts, and relays) and was by no means a fully electronic machine. The two men began working as a team and were ideally suited to each other. Mauchly was people-oriented and a visionary, while Eckert was a focused, detail-oriented person. Each man's mind and personality complemented the other's.

J. Presper Eckert

Invents first general-purpose electronic computer

Ever since 1941 when it entered World War II (1939–45), the United States had been trying to develop a machine that would perform all sorts of war-related calculations automatically and rapidly. Like many other institutions, the Moore School of Electrical Engineering was bidding for a contract with the government to develop such a machine. In 1942, Eckert and Mauchly submitted a proposal to the U.S. Army for a machine that would compute one thousand times faster than any existing device. On June 5, 1943, the Moore School signed a contract with the Army's Ordnance Department to build what came to be called the ENIAC (Electronic Numerical Integrator and Computer). When they first began, Eckert was the project's chief engineer and the only full-time person. Mauchly served as consultant and eventually found himself heading a team of fifty people.

Eckert's main job was designing the machine's electronic circuits, and his basic strategy in doing this was to adapt existing technology whenever possible. Wartime research was always rushed and his work was no exception. This meant that he would use nearly eighteen thousand vacuum tubes in order to make the machine fully electronic. Vacuum tubes are sealed glass tubes that looked like light bulbs, and they were used in all radios of the time, and eventually would come to power early television sets. In an electronic circuit, vacuum tubes work like a valve and either permit or stop an electric current from flowing (they were either on or off). Although vacuum tubes were the best technology available at the time (they would eventually be replaced by transistors, which would be replaced by chip technology), they presented Eckert with many problems. Vacuum tubes were usually very large, always fragile, used a great deal of power, and had a limited life. Adding to these problems was the fact that if one of the nearly eighteen thousand vacuum tubes burned out, the machine would stop performing its calculations. Eckert, however, was an expert on electric organs, which used vacuum tubes, and therefore was very familiar with how to best manage them. He knew, for example, that vacuum tubes could be made to last a very long time if they were run at low power. In this and every aspect of his work on ENIAC, Eckert stressed simplicity and careful standards, and he and Mauchly's team were able to accomplish a great deal in a fairly short time.

What the team eventually produced was a truly gigantic machine. Weighing over 30 tons, it was 80 feet long, 8 feet high, 3 feet deep, and took up 1,800 square feet of floor space. It was the size of a very large room. When it was formally dedicated in February 1946, it successfully completed its first problem, a secret calculation for America's future hydrogen bomb, in twenty seconds. This problem would have taken existing calculating machines forty hours to compute. Although the ENIAC was an enormous and power-hungry machine that was slow compared to twenty-first century computers, its calculating speed was one thousand times faster than anything else available in 1946. It was capable of performing five thousand additions per second. The ENIAC impressed everyone with its all-purpose ability to perform **addition** (see entry in volume 1), **subtraction** (see entry in volume 2), **multiplication** (see entry in volume 2), and **division** problems; compare quantities; and extract **square roots** (see entry in volume 2). Although the ENIAC did not become operational until after World War II had ended, it was put to use by the Army and was later used in the development of the hydrogen bomb. Eckert and Mauchly had produced the first general-purpose electronic digital computer, and it would influence all other electronic computers that would follow.

EDVAC, the improved model

While designing and building the very successful ENIAC, Eckert and Mauchly realized that despite its potential and actual accomplishments, the ENIAC had many weaknesses. This led them to begin work on an improved model even before the ENIAC was finished. One of ENIAC's biggest problems was that it had no real memory, and when Hungarian American mathematician **John von Neumann** (1903–1957; see entry in volume 2) came to the Moore School as a consultant, he joined Mauchly and Eckert in working toward building a computer that would store both its program and its data in memory. By 1951, they produced such a machine. Known as the EDVAC (Electronic Discrete Variable Automatic Computer), this improved machine used only 3,600 vacuum tubes, had its own internal memory for storing programs, and took up a mere 490 square feet.

J. Presper
Eckert

With the war over and the ENIAC a success, Eckert and Mauchly felt that they should be allowed to obtain the patent rights to their new machine. As they began this process, the University of Pennsylvania decided that it would retain future patents on all projects that were developed by its employees. This meant that Eckert and Mauchly would have to surrender any rights to their future work, so the two decided to resign in March 1946.

Develops the first commercial computer

Eckert and Mauchly proceeded to form their own company in Philadelphia, the Electronic Control Company (later called Eckert-Mauchly Computer Corporation), despite the fact that IBM had offered Eckert a job and his own lab. The two immediately obtained contracts from two government agencies—the National Bureau of Standards (NBS) and the Census Bureau—and with the cash these brought in, kept going until they obtained a major contract from NBS to build what they called the Universal Automatic Computer (UNIVAC). This was to be a major advance in computing. While they worked on this ambitious project, they signed a contract with Northrop Aircraft Company to build a small computer for navigating airplanes, which came to be called the BINAC (Binary Automatic Computer). Completed in August 1949, the BINAC offered a major advance since its data was stored on magnetic tape rather than on punched cards.

For Eckert and Mauchly, however, the BINAC was just a stepping-stone toward their real ambition of building a computer for the everyday business world and not just for the military. This would be the UNIVAC, and eventually they would build forty-six of them. Like the BINAC, the UNIVAC also used magnetic tape for data storage. It was much smaller than the ENIAC and made of fewer parts. It also had an internal memory for storing programs and could be accessed by typewriter keyboards. One of its major advances was its ability to handle both numerical and alphabetical information. When they delivered the first UNIVAC to the Census Bureau in March 1951, the two men achieved their goal of producing the world's first general-purpose commercial computer that was able to handle a wide variety of applications. One of its better-known uses was demonstrated during the 1952 presidential election when it predicted correctly, less than one hour after the polls had closed, that

A man stands behind the ENIAC
computer, a machine devised by
J. Presper Eckert and John
Mauchly.
*Reproduced by permission of the
Corbis Corporation.*

Dwight D. Eisenhower (1890–1969) would defeat Adlai E. Stevenson (1900–1965) and become the next U.S. president.

Financial and legal problems

Despite their technical success with the UNIVAC, Eckert and
Mauchly suffered financial problems running their new company.
Although they were probably better at designing computers than running a business, one of their major problems was that everything they

did was so new and untried that it was extremely difficult to estimate production costs. These financial problems led them to sell their company to the Remington Rand Corporation in 1950. It was as part of Remington Rand that UNIVAC was completed. Although both Eckert and Mauchly became employees of Remington Rand (which later became Sperry Rand and then Unisys), they split as a team in 1959. Eckert remained at the company and eventually became director of its UNIVAC Division. Mauchly went on to start his own company, then moved to Dynatrend Inc. and Marketrend Inc.

Years later, Eckert and Mauchly's patent on the ENIAC was challenged in court, and a controversy emerged over who actually invented the first electronic computer. After a long legal battle, a court ruled in 1974 that Iowa State University physics professor John Vincent Atanasoff (1903–1995) actually built a model of the first electronic digital computer in 1942. Although this machine never became operational, the court declared Atanasoff and his colleague, Clifford E. Berry (1918–1963), as the inventors of the modern computer. Despite this ruling, there is no way to ignore the fact that Eckert and Mauchly produced the first real, workable, electronic digital computer that was able to solve a variety of real-world problems with far greater speed than anything else. It then led to the EDVAC, the BINAC, and ultimately, the highly successful commercial computer, the UNIVAC, which helped launch the modern age of the computer. In fact, in the early 1950s, the word "UNIVAC" was used in the same way "computer" is used today.

By the time Eckert retired from Unisys in 1989, he had resigned himself to the court's decision. The honors and awards he received during his lifetime indicate how the majority of historians feel about his impact. In 1969, Eckert was awarded the National Medal of Science, the nation's highest award for distinguished achievement in science, mathematics, and engineering. Eckert died on June 3, 1995, after a prolonged battle with leukemia (a blood disease). He was survived by his second wife, Judith, with whom he had two children, Laura and Gregory. He also had two sons, John Presper III and Christopher, with his first wife, Hester Caldwell, who died in 1952.

For More Information

Cortada, James W. *Historical Dictionary of Data Processing: Biographies.* New York: Greenwood Press, 1987.

Mathematicians and Computer Wizards. Detroit: Macmillan USA, 2001.

O'Connor, J. J., and E. F. Robertson. "John Presper Eckert." *The MacTutor History of Mathematics Archive.* School of Mathematics and Statistics, University of St. Andrews, Scotland. http://www.groups.dcs.st-andrews.ac.uk/~history/Mathematicians/Eckert_John.html (accessed June 24, 2002).

Slater, Robert. *Portraits in Silicon.* Cambridge, MA: MIT Press, 1987.

Vogt, Peter. "Presper Eckert Interview." *National Museum of American History, Smithsonian Institution, Computer History Collection.* http://americanhistory.si.edu/csr/comphist/eckert.htm (accessed June 24, 2002).

J. Presper Eckert

Born January 1, 1878
Lonborg, Jutland, Denmark

Died February 3, 1929
Copenhagen, Denmark

Danish probabilitist

Agner K. Erlang

Agner K. Erlang was a pioneer in the study of telecommunications traffic or telephone networks. He worked out formulas that gave the **probability** (see entry in volume 2) that a user will get a busy signal instead of a dial tone, and the time a user would have to wait when placing a call in a system that can hold calls. This information is then used to calculate the number of circuits needed to give a certain level of service. The mathematics underlying today's complex telephone networks is still based on his work.

Educated at father's school

Agner Krarup Erlang was born on New Year's Day in 1878 at Lonborg, near Tarm in Jutland, on the mainland of Denmark. His father, Hans Nielsen Erlang, was the village schoolmaster and parish clerk. His mother, Magdalene Krarup, came from a long line of clergymen, and among her ancestors was Danish mathematician Thomas Fincke (1561–1656). Erlang had one older brother, Frederik, and two younger sisters, Marie and Ingeborg. As a youngster, Erlang attended his father's school. He is said to have spent many a playful evening reading with his brother. One of young Agner's favorite subjects was astronomy, and he even wrote poems on astronomical subjects. At home, his father and an assistant teacher tutored him to prepare him for his preliminary examination. Although Erlang took and passed this test with distinction, he needed special permission to do so since he was only fourteen years old at the time.

After passing this exam, Erlang began teaching at his father's school and continued his studies to prepare for the entrance examination to the University of Copenhagen. During this time, he also learned French and Latin. He soon moved to the city of Hillerød where he stayed with a family relative. All of his hard work paid off in 1896 when he passed his exam, again with distinction, and won a scholarship to the University of Copenhagen. A scholarship was essential since his parents were not well off and Erlang could not afford to pay the full tuition. At the university, Erlang was a good student and graduated in 1901, majoring in mathematics. He also took physics, astronomy, and chemistry as his secondary subjects.

Chance meeting decides his fate

Over the next seven years, Erlang taught at various schools. Despite his love of scientific research, he found that he was a natural teacher who enjoyed the classroom environment. Although he was a good teacher, he was not a very social person. His personal style was to say very little, and hang back and observe rather than be closely involved with things or people. He did have many friends, however, and they jokingly nicknamed him "the private person." Erlang was not all work, and often spent his summer holidays traveling to France, Sweden, Germany, and England, usually visiting their art galleries and museums. Although he was a full-time teacher, he continued to study mathematics and do research. It was during these teaching years that he received an award in 1904 for an essay he submitted to the University of Copenhagen concerning a solution to a long-standing problem put forth by Dutch astronomer and mathematical physicist **Christiaan Huygens** (1629–1695; see entry in volume 1). Erlang's mathematical interests soon turned toward the theory of probability (the mathematical study of the likeliness of an event occurring), and he decided to keep up his mathematical interests by joining the Danish Mathematics Association. It was through his contact with this group that he made an acquaintance who would change his life.

Agner K. Erlang

One of the members of the association, Johan Jensen (1859–1925), was also chief engineer at the Copenhagen Telephone Company, and after meeting Erlang and learning of his interests in probability, Jensen persuaded him to leave teaching and join the telephone company. Jensen was convinced that the company needed someone with Erlang's skills to help solve the problems it was having with long waiting times for telephone calls. Once the company's managing director met and spoke with Erlang, he offered to hire him as "scientific collaborator" and make him the leader of his own laboratory. Erlang accepted and joined the company in 1908.

Applies probability to telephone system

Erlang studied the company's phone situation and applied his mathematical knowledge of probability to the different problems the company was having providing good, fast phone service to its customers. Within a year, he published his first work on this problem, "The Theory of Probabilities and Telephone Conversations." In this important work, he used a single circuit in a single village as a model, and developed a formula that demonstrated the number of calls to arrive during a certain period of time followed a known value, called Poisson's (pronounced pwah-SOHN) law of distribution. Using this formula, the company could calculate the **fraction** (see entry in volume 1) of callers who would have to wait when trying to place a call to someone outside the village, which in turn, allowed the company to provide more efficient service.

Erlang's 1909 paper was the first of its kind, and it gained him worldwide recognition as his formula was accepted and eventually used by the British Post Office. Erlang continued his work, and in 1917 produced another important paper in which he offered a formula for the probability of barred access or a busy signal for a group of circuits. As Erlang continued to publish and lecture on his specialty, his work became even better known, and several of his papers were translated into English, French, and German. His work was not always the easiest to understand, however, since he wrote as he spoke, very briefly, and sometimes did not include any proofs. Still, his work on the theory of telephone traffic became universally accepted and was used by all major telephone companies throughout the world. Further, the beauty of Erlang's formulas is that they can be applied to any system with a limited number of servers that has customers who request service at random times.

Erna Schneider Hoover

If Agner Erlang's formulas allowed early telephone switching systems to become truly operational, it was the computerized telephone switching system invented by American mathematical researcher Erna Schneider Hoover (1926–) that enabled the telephone to become an essential part of the modern world.

Hoover was not a typical research worker at Bell Labs. First, she was a woman in a male-dominated field. She also had a varied educational background: she received her bachelor's degree in medieval history from Wellesley College and earned a Ph.D. in philosophy and foundations of mathematics from Yale University.

When Hoover began work at Bell Labs in New Jersey in 1954, telecommunications was in the process of a radical change. In 1947, the transistor was invented and this new device had offered the potential of becoming a key part of all electronic telephone switching systems. It was Hoover, however, who would make it so. While in the hospital after giving birth to one of her three daughters, Hoover drew up the first sketches of what would be the first computerized switching system for telephone call traffic. With the growth in domestic telephone traffic after World War II (1939–45), telephone systems were being overwhelmed with the increasing number of calls coming through, since the old, hard-wired mechanical switching equipment (now seen in old photos of operators plugging lines into big phone boards) were not up to the job. Hoover's solution was to use a computer to monitor the frequency of incoming calls at different times, and to then adjust the call acceptance rate as necessary. Hoover's system eliminated the danger of overloading, and she was granted one of the first software patents ever issued. Bell Labs then made Hoover the first woman supervisor of one of its technical departments. Hoover's work led to stored program control (SPC) systems that are still used today in business, public communications, and even on the Internet.

Agner K. Erlang

The *Erlang* as a unit of measurement

In addition to Erlang's widely used formulas, his name was adopted internationally in 1946, and the *Erlang* became an accepted unit of telecommunication traffic measurement. An Erlang unit is the total traffic volume of one hour. More recently, ERLANG also was chosen as the name for a new programming language developed by a team of employees at the giant Swedish communications firm Ericsson. This language is used to write huge, real-time control programs for telephone exchanges and network switches.

Erlang lived a quiet life in Denmark, despite his international reputation. He never married and lived with his sister, usually devoting most of his time and energy to his work and studies. He was a book collector and was known in his country as a charitable man who would help anyone in need. When his sister founded a home for mentally ill women, Erlang was one of its biggest supporters. Although he worked for the Copenhagen Telephone Company for nearly twenty years and never had to take time off for illness, he suddenly was hospitalized during January 1929, and after an abdominal operation, died that same year at the age of fifty-one.

For More Information

"Agner Krarup Erlang (1878–1929)." *Millenium Mathematics Project.* http://plus.maths.org/issue2/erlang/ (accessed June 25, 2002).

The Biographical Dictionary of Scientists. 2d ed. New York: Oxford University Press, 1994.

Brown, David E. *Inventing Modern America: From the Microwave to the Mouse.* Cambridge, MA: MIT Press, 2002.

Kei, Choi Kin. *The Life and Times of A. K. Erlang.* http://www.eie.polyu.edu.hk/~ckleung/tcnet/ss/choikk/ (accessed June 25, 2002).

O'Connor, J. J., and E. F. Robertson. "Agner Krarup Erlang." *The MacTutor History of Mathematics Archive.* School of Mathematics and Statistics, University of St. Andrews, Scotland. http://www.groups.dcs.st-andrews.ac.uk/~history/Mathematicians/Erlang.html (accessed June 25, 2002).

Born April 8, 1939
Hastings, Nebraska

American mathematical statistician and educator

Mary Gray

Mary Gray is best known as one of the founders and the first president of the Association for Women in Mathematics. Although her specialty is mathematics, she has worked in the field of physics and also has a doctor of laws degree. She has served as chair of the department of mathematics and **statistics** (see entry in volume 2) at American University since 1977, and became the first woman elected vice president of the American Mathematical Society.

Top of the class

As the only child of native Nebraskans, Mary Wheat Gray received all the attention she needed. Her father, Neil C. Wheat, who worked as a policeman, truck driver, mechanic, and traffic manager, read to his young daughter a great deal and was already challenging her with mental arithmetic when she was only five years old. Her mother, Lillie W. Alves, who had been a school-teacher before marriage, gave Mary a sense of history by telling her stories about boarding with a family when she taught Native American children in a one-room schoolhouse. She recalled how the Native American families had set up camp in a nearby field so that their children could attend school. In high school, Mary's

Mary Gray

favorite subjects were mathematics, history, and physics, and she excelled in everything she did.

Choosing to remain in Hastings for college, Gray attended Hastings College and went on to receive her bachelor's degree in mathematics and physics in 1959 at the age of twenty. She was ranked at the top of her graduating class and received a Fulbright scholarship after graduating. This allowed her to study abroad for a year; she chose to attend the J. W. Goethe (pronounced GER-tuh) Institute in Frankfurt, Germany.

Decides on a career in mathematics

Upon returning to the United States, Gray entered graduate school at the University of Kansas and decided upon mathematics as her specialty. Although she had considered going to law school, she decided to take advantage of the many new scholarship programs offered by the government to students who would specialize in math or science. Only a few years before, the Soviet Union had shocked the world by launching *Sputnik,* the world's first artificial satellite, and the United States felt this was a sign that it had fallen behind in educating its young people in science and math. Gray received a shock, however, when the instructor of her first graduate school class asked her bluntly, "What are you doing here? Why don't you stay home and take care of kids?" This was her first encounter with something she would meet throughout her career—gender discrimination in mathematics. In the face of such difficulties, however, Mary found that she became only more determined.

In 1962, she was awarded her master's degree in mathematics from Kansas and was hired there as an assistant instructor. Still in need of financial help, she received fellowships (grants) from the National Science Foundation and the National Defense Education Act, and received her Ph.D. in mathematics from the University of Kansas in 1964. She was only the second woman to earn a Ph.D. in mathematics at that school; the first woman received her doctorate in 1926. Taking her doctorate and moving to the west coast, she taught for one year at the University of California, Berkeley, and then took an assistant professorship at California State University at Hayward. By 1968, she had become an associate professor of mathematics, and accepted a teaching position at American University in Washington, D.C. While at California

State, she had met a faculty member, Alfred Gray, whom she married in 1964. He was also a mathematician, and when they moved to the Washington suburbs in 1968, he took a teaching position at nearby University of Maryland.

Concern for social justice

Throughout her life, Gray has been an activist when it comes to issues that are important to her. She has never been one to sit by quietly when she sees a wrong being done. When she was a student at Kansas, she organized a successful protest against a local grocery store that would not hire Native Americans as grocery baggers. When she later changed roles from student to professor, she did not change her attitude or actions and worked to support the cause of farm workers in California. And with her experience as a woman in a field dominated by men, she became increasingly concerned with the rights and fate of her female colleagues.

In 1971, when Gray was a full professor at American University but before she became the department chair, she attended a meeting of the male-dominated American Mathematical Society (AMS) held in Atlantic City, New Jersey. When she attempted to attend a meeting of a particular council, she was denied permission. Gray stood her ground and argued that the AMS bylaws did not say that any meetings were closed to women. When told that everyone always understood this and that all were following a sort of "gentlemen's agreement," Gray stated, "Well obviously I'm no gentleman." Following this head-on encounter, she and many other women were allowed to attend such meetings.

Founds Association for Women in Mathematics

To have an alternative from the male-run AMS, Gray helped create the Association for Women in Mathematics (AWM). It was Gray who set the goals and the agenda for her new organization. Her main goal was to assist women mathematicians professionally, and to encourage more women and girls to study mathematics. In a 1991 article on the history of the AWM by Lenore Blum, Gray is described as "the mother of us all." In many ways, this is not an exaggeration, for were it not for her vision, determination, and hard work, there might never have been an organization of and for women in mathematics. This point is underscored by for-

Mary Gray

mer AWM president Carol Wood who stated in the AWM article, "My overwhelming sense … is that AWM would not have existed when it did, if at all, without the energy and vision of Mary Gray." Gray served as AWM president from 1971 to 1973.

Gray worked as a physicist at the National Bureau of Standards (now the National Institute of Science and Technology) during graduate school, and much later used her skills in statistics to study how they were used or misused in cases involving human rights issues. But Gray's activism and concern for the careers of women mathematicians necessarily absorbed much of her time, and as she became more of an advocate for gender equity or fairness, she spent less time with mathematics itself. For example, the AWM, which she helped found, took an active role in promoting the rights of Soviet and Chinese women mathematicians and tried to help them emigrate (leave their country).

Obtains law degree

As an activist always in the forefront of women's issues and the leader of a growing association, Gray found herself testifying in court more and more. This led her to return to school to learn more about the law, and in 1979 she received a doctoral degree in law from American University's Washington College of Law. Since then, her academic interests broadened from algebra and statistics to computer law. The degree also enabled her to become even more active in promoting and defending the rights of women throughout the world. She is regularly called upon by congressional committees to discuss such subjects as mathematical education, age and gender discrimination, and many other issues related to the role of women in the sciences.

Gray has gone on to involve herself in an ever-broadening way with many organizations that are concerned in some manner with the role of women. She is a member of several important organizations, such as Amnesty International, the American Civil Liberties Union, and the American–Middle East Education Foundation, and has held leadership positions in many of them. She also has spoken on the important role and use of statistics in such varied countries as China, England, Morocco, and Jordan. In 1994, she received the Mentor Award for Lifetime Achievement from the American Association for the Advancement of Science. As noted in *Biographies of Women Mathematicians*, this award specifically recognized "the extraordinary number of women and underrepresented minorities she has affected in her career both directly and indirectly through the influence of her former students and the programs she initiated and developed." In 2001, she and nine others were the recipients of the Presidential Award for Excellence in Science, Mathematics, and Engineering Mentoring.

Away from work, Gary enjoys traveling extensively. She usually manages to locate the nearest opera house wherever she visits in order to indulge her deep love of opera, an interest shared by her husband, Alfred, until his death in 1998. Mary Gray is a shining example not only of how one person can make a great difference, but also of the fact that mathematics exists in a very real world that is sometimes not always fair or just. Gray, however, has dedicated her life not only to making sure that she does her utmost to improve this situation, but to the cause of making sure that mathematics is both properly understood and properly used.

For More Information

American Men and Women of Science. New York: R. R. Bowker, 1989–98.

Blum, Lenore. "Mary Gray (1971–1973): The Mother of Us All." *A Brief History of the Association for Women in Mathematics: The Presidents' Perspectives.* http://www.awm-math.org/articles/notices/199107/blum/node2.html (accessed June 25, 2002).

Kelly, Virginia Myers. "Mary Gray Wins Presidential Award." *News at AU.* http://veracity.univpubs.american.edu/weekly-past/ 121101/story_2.html (accessed June 25, 2002).

Morrow, Charlene, and Teri Perl, eds. *Notable Women in Mathematics: A Biographical Dictionary.* Westport, CT: Greenwood Press, 1998, pp. 71–76.

"President Bush to Honor American University Professor Mary Gray, Dec. 12, at White House Ceremony Honoring Science, Mathematics and Engineering Mentors." *American University News.* http://domino.american.edu/AU/media/mediarel.nsf/608575dac58ec4a785256869007c9cba/57953 cdfb387525085256b19005a9425?OpenDocument (accessed June 25, 2002).

Mary Gray

Mary Gray

Riddle, Larry. "Mary Gray." *Biographies of Women Mathematicians.* http://www.agnesscott.edu/lriddle/women/gray.htm (accessed June 25, 2002).

Born December 8, 1865
Versailles, France

Died October 17, 1963
Paris, France

French analyst

Jacques-Salomon Hadamard

Jacques-Salomon Hadamard is widely considered the most important French mathematician of the twentieth century. Called the "living legend of mathematics," he was a true creative genius whose encyclopedic knowledge enabled him to have an impact on nearly every branch of mathematics.

Jacques-Salomon Hadamard.
Reproduced by permission of
Corbis-Bettmann.

A family of teachers

Jacques-Salomon Hadamard (pronounced odd-uh-MARR) was the only son of Amédée Hadamard, a teacher of Latin, grammar, classics, history, and geography at a prominent high school in Paris, and Claire Marie Jeanne Picard, an accomplished pianist and distinguished music teacher who gave lessons in the family home. His two younger sisters, Jeanne and Suzanne, both died before they reached five years old. When Hadamard was born, his father was teaching at Lycée Imperial in the French city of Versailles (pronounced vehr-SYE), but the family moved to Paris when Jacques was three as his father accepted a teaching position at the Lycée Charlemagne. Soon after moving, the Franco-Prussian War (1870–71) broke out and Paris came under attack. The Germans blockaded the city, allowing no one to enter or leave,

Jacques-Salomon Hadamard

and the Parisians were soon forced to eat their horses as well as cats and dogs as things got more desperate. The Hadamard family was forced to eat elephant meat to survive. After the city surrendered in January 1871, a civil war broke out in Paris during which the Hadamards' home was burned to the ground.

With the war over and a treaty signed, life returned to normal and young Hadamard attended the Lycée Charlemagne where his father taught. During his first few years at school, he showed ability in all of his subjects except mathematics. As noted in *The Mac-Tutor History of Mathematics Archive,* years later, Hadamard recalled, "In arithmetic, until the fifth grade, I was last or nearly last." This did not go on very long, however, and he was soon winning prizes and mathematical competitions by the sixth grade. When his father was transferred to the Lycée Louis-le-Grand, his son also moved to this school, and it was from there that he graduated in 1882. The next year, he showed his true ability by winning first prize in both **algebra** (see entry in volume 1) and mechanics in the Concours Général, a national competition for high school students. In 1884, Hadamard placed first in the entrance examinations for two of the best schools, École Polytechnique and École Normale Supérieure. He chose to attend École Normale Supérieure, and it was there that he became a real mathematician, graduating in 1888.

Embarks on a distinguished career

After graduating, Hadamard taught briefly at two Paris schools before accepting a position at the Lycée Buffon in 1890. He would remain there for three years, during which time he also was working toward his doctorate degree. The year 1892 was a significant one for Hadamard. In this year, he received his doctorate, and in the same year also was awarded the Grand Prix, the highest honor the French Academy of Sciences could bestow. This award was given on the basis of Hadamard's thesis, which was considered revolutionary by the French Academy. This work was in the difficult field called analytic functions, and Hadamard was the first to solve certain problems in that field that had always been considered unsolvable. In this same year, Hadamard married Louise-Anna Trenel, with whom he would have five children. The two had known each other from childhood, and both shared a deep love of music. Also in 1892, the newlyweds moved to Bordeaux (pronounced boar-DOE), France, after he accepted a position at the Faculté des Sciences. By 1896, Hadamard had become professor of astronomy and rational mechanics at Faculté des Sciences.

Hadamard would spend fours years in Bordeaux, and it was a very productive time for him. During these years, he published twenty-nine papers, which were remarkable not only in their depth but in the broad range of topics they covered. The ability to deal in-depth with almost any branch of mathematics would become a trait for which he would be well known. Already, Hadamard was well on his way to earning his other nickname, "the universal mathematician." It was during this time that he solved a long-standing problem that had defied some of the world's best mathematicians. While many had offered or suggested what is called the **prime number** (see entry in volume 2) theorem, no one had ever been able to prove it. A prime number is a natural or counting number greater than 1 that has only two factors—itself and 1. A factor of a number divides the number evenly and has no remainder. A prime number can also be described as a natural number that cannot be made by multiplying smaller numbers together. For example, the numbers 2, 3, 5, 7, 11, 13, and 17 (and so on) are prime numbers. As far back as the third century B.C.E., the Greeks had suggested, but not proven, a prime number theorem that stated there was an infinite (never-ending) number of primes. In 1896, Hadamard used powerful new methods and became the first to offer a correct proof. This achievement alone would have been enough to rank him with the top mathematicians, yet he lived a long and productive life in which he made regular and often spectacular contributions.

Politics and math

During his time in Bordeaux, Hadamard became involved in what became known as the Dreyfus affair. Alfred Dreyfus (1859–1935) was a French army officer who was wrongly accused and convicted of selling military secrets to the Germans. Dreyfus, who was Jewish, was a distant relative of Hadamard's wife, and once Hadamard came to understand the facts and circumstances, he believed that Dreyfus had been a victim not only of a conspiracy but of religious discrimination as well. When Hadamard moved back to Paris in 1897 to teach at the Sorbonne, he learned that the evidence against

Jacques-Salomon Hadamard

The cover of the January 13, 1895, issue of *Le Petit Journal* portrays Captain Alfred Dreyfus (standing at attention as a soldier breaks his sword) as a traitor. Dreyfus was eventually cleared of all charges.
Reproduced by permission of Moon Mountain Publishing, Inc.

Dreyfus had been forged, and he immediately became a leading member of those who were trying to save Dreyfus from a life sentence in prison. For years, all of France was split into two camps on the Dreyfus Affair, and Hadamard took an active part in trying to clear Dreyfus's name. This finally happened in 1906 as a retrial reversed his conviction and even decorated Dreyfus.

During this time in Paris, Hadamard was also teaching and working on mathematics, and it was at this time that his interest turned more toward mathematical physics. Hadamard had always been critical of mathematicians who stayed in one narrow field and never investigated anything beyond their specialty. He, on the other hand, always looked for connections between mathematics and other fields. Because of his genius and wide-ranging interests, he had a major impact on several branches of mathematics and influenced several other very applied fields. These included topology (the study of the properties of geometric figures that are not changed by stretching or bending), hydrodynamics (the study of fluids in motion), mechanics (the study of the action of forces), **probability** (see entry in volume 2) theory (the study of the likeliness of an event occurring), and even **logic** (see entry in volume 2). After leaving the Sorbonne in 1909, Hadamard went on to accept increasingly prestigious and sometimes overlapping positions, such as at the Collège de France (1909–37), the École Polytechnique (1912–37), and the École Centrale des Arts et Manufactures (1920–37). He was elected to the Académie des Sciences in 1912 as well as several foreign academies, and was also elected president of the French Mathematical Society.

Personal tragedies

World War I (1914–18) led to a great sadness for Hadamard as his two older sons were killed in the war. First Pierre was killed at Verdun in 1916, and then Etienne was killed nearby only two months later. For years afterward, Hadamard worked at his mathematics as long and as hard as he could as a way of dealing with his loss and pain. Sadly, the situation would repeat itself during the next major conflict, World War II (1939–45), when Hadamard would again receive the terrible news that another beloved son, Matthieu, had been killed in action in 1944. Hadamard was in the United States at this time since he and his family had left France in 1940 when the Germans captured it. Although Hadamard had retired from teaching in 1937, he accepted a visiting professorship at Columbia University for the war years, but then went to England for a year after his third son died. With the war's end, Hadamard returned to Paris as soon as he could and became an active campaigner for peace among nations. Finally, as a very old man of ninety-six, Hadamard endured yet another tragedy when his grandson, Etienne, was killed in a mountaineering accident. Too old to lose himself in work, Hadamard never left his house and seemed to want his end to come. He died in Paris at the age of nintey-seven.

Jacques-Salomon Hadamard

Jacques-Salomon Hadamard

For More Information

Abbott, David, ed. *Mathematicians.* New York: Peter Bedrick Books, 1986.

Hadamard, Jacques. *The Mathematician's Mind: The Psychology of Invention in the Mathematical Field.* Princeton, NJ: Princeton University Press, 1996.

Mandelbrot, S. "Jacques Hadamard." In *Biographical Dictionary of Mathematicians.* New York: Charles Scribner's Sons, 1991, pp. 958–60.

Maz'ia, Vladimir G., and Tatyana Shaposhnikova. *Jacques Hadamard: A Universal Mathematician.* Providence, RI: American Mathematical Society, 1998.

O'Connor, J. J., and E. F. Robertson. "Jacques Salomon Hadamard." *The MacTutor History of Mathematics Archive.* School of Mathematics and Statistics, University of St. Andrews, Scotland. http://www.groups.dcs.st-andrews.ac.uk/~history/Mathematicians/Hadamard.html (accessed June 25, 2002).

Born 1560
Oxfordshire, England

Died July 2, 1621
London, England

English algebraist and astronomer

Thomas Harriot

Thomas Harriot was a major English mathematician who pioneered the use of symbolism in **algebra** (see entry in volume 1) and was responsible for a number of new mathematical ideas. Considered a "universal genius" by his peers, he displayed accomplished work in astronomy and optics and would have been even more influential had he published his work.

Little known of early years

Thomas Harriot was born in Oxfordshire, England, in 1560. Historians know about his beginnings because his name, age, and place of birth are entered in the records of St. Mary's Hall, at Oxford University, for 1577. His parents's names are not given, although the record describes his father as a commoner. Despite the fact that his family is described as among the common people and the university lowered its tuition because of this, his parents must not have been extremely poor or he would never have had the opportunity to be admitted to such a prestigious institution. There is some evidence that his father may have been a blacksmith. Harriot lived at St. Mary's Hall until he graduated with a bachelor of arts degree in 1580. Besides studying mathematics

Thomas Harriot

and science, he took full advantage of his time at Oxford and studied other subjects that were beyond the normal requirements. Scholars know that Harriot was fluent in Greek and Latin and was very familiar with the **geometry** (see entry in volume 1) of Greek mathematician **Euclid of Alexandria** (see entry in volume 1) and the major schools of natural philosophy (science). Little is known of his personality except for the fact that he apparently preferred the student custom of wearing black and continued to do so throughout his life.

Sir Walter Raleigh his patron

Upon graduating from Oxford in 1580, Harriot went to London and began giving private lessons in mathematics. His name first appears in 1583 working as an assistant to explorer Sir Walter Raleigh (1554–1618). Raleigh had become a favorite of Queen Elizabeth I (1533–1603), and was made captain of the queen's guard in 1587. Raleigh became Harriot's patron (someone who uses his money or influence to help another individual) so that Harriot could train Raleigh's sea captains in navigation. England was about to challenge Portugal and Spain in exploring and colonizing the New World, and Raleigh had obtained permission to begin an expedition. Harriot was hired to provide navigational instructions and scientific advice to Raleigh's seamen, and he taught a very detailed and thorough course in London while living in Raleigh's home.

In 1585, Harriot was given the chance to test his navigational abilities when he joined a Raleigh-sponsored expedition to America as its scientific observer. He sailed to Virginia and then on to a colony, in what is now North Carolina, where he remained for about a year. While there, he cataloged the new plants and animals he encountered, learned how to smoke tobacco, and recorded what the colonists had to do on a daily basis to survive. Three years later, his findings were published in *A Briefe and True Report of the New Found Land of Virginia*. The short book soon became a popular instructional guide for future colonists. Upon his return, Harriot lived for the next ten years on an estate in Ireland, which Raleigh gave him. During this time, he continued to work for Raleigh and helped with his various business affairs.

In 1592, Raleigh suddenly fell out of favor with the queen, due to an unpopular secret marriage. Associates of his also came to suffer.

Thomas Harriot

British explorer Sir Walter Raleigh, for whom Thomas Harriot worked. *Public domain.*

For years, Raleigh went through a series of accusations, arrests, trials, imprisonments, and release until he was finally executed in 1618 on the charge of conspiring against King James I (1566–1625), who had ascended to the throne upon Elizabeth's death in 1603.

Harriot naturally was involved in Raleigh's misfortune, and was even imprisoned at one point. Long before Raleigh's death, however, Harriot began working for Henry Percy (1564–1632), the ninth earl of Northumberland, in 1595, and he moved into the

Harriot Goes to the New World

Many historians know of Thomas Harriot, not because of his mathematical or scientific accomplishments, but because of his association with Sir Walter Raleigh, tobacco, and the Lost Colony of Roanoke. Harriot was sent to the New World in the spring of 1585 by Sir Walter Raleigh and served as the expedition's scientific advisor. Raleigh later used Harriot's fascinating book, *A Briefe and True Report of the New Found Land of Virginia,* which contained Harriot's observations, to promote further colonization (settlement).

Although tobacco had already been introduced to Europe by 1588 when Harriot's book was first published, Harriot wrote about having learned to "drink" tobacco smoke from the Indians, which he said was a cure for many ailments. It would not be long before using tobacco become popular in Europe; soon, tobacco cultivation was an important part of the economy of colonial America as well. In fact, tobacco became the chief commodity (main product) that colonialists used when trading for European manufactured articles.

While in the New World, Harriot was based on Roanoke Island, in what would become known as the famous "Lost Colony." It was there that Virginia Dare, the first child of English parents to be born in the New World, was born on August 18, 1587. Harriot was gone by that time, and by 1590, all the settlers had vanished, leaving no real clues to what caused them to totally disappear. Today there is a national historic site located on Roanoke Island.

earl's estate. There, Percy, who was known as the "wizard earl" because of his involvement with science, gave Harriot his own laboratory and the freedom to conduct whatever scientific experiments he chose. During this time, Harriot conducted investigations in astronomy and optics (the study of light). He is credited with being the first person in England to build and use a telescope, which he used to draw some of the earliest maps of the moon's surface. He observed Jupiter's moons, studied sunspots, and calculated the speed of the sun's rotation. He also studied light and noticed the way it was bent when it passed through a glass or through liquid—thereby discovering the law of refraction (the bending of light) twenty years before Dutch mathematician Willebrord Snell (1591–1626). Harriot also considered many practical problems and studied the flight of cannon shells in order to improve their accuracy. He even investigated specific gravity (the density of a substance compared to the density of water) and developed tables of the specific gravity of various materials. Unfortunately, he never published any of his findings. During these years, he did not neglect his former patron, Raleigh, and was even present to record Raleigh's last words before he was executed.

Love of algebra

Throughout his life and despite all his varied interests and studies, Harriot's great love was always algebra. Today, he is considered the founder of the English school of algebra, although during his lifetime, he was not very influential because he published nothing. His only mathematical publication, *Artis Analyticae Praxis ad Aequationes Algebraicas Resolvendas* (The Analytical Arts Applied to Solving Algebraic Equations), came ten years after his death and is considered to be one of the most important books on algebra. This work contained a number of Harriot's innovations (new ideas and methods) that would become standard practice. His use of symbolism was especially important, and he also included the use of small letters of the alphabet for unknowns. Further, his use of the Greek cross "+" for **addition** (see entry in volume 1) and "=" for equality—both offered first by English mathematician **Robert Recorde** (c. 1510–1558)—were very influential and mostly responsible for their acceptance and use in England. As someone who always tried to make things simple in algebra, he is credited with inventing the signs for "greater than" (>) and "less than" (<). Harriot applied his simplified notation systems to other areas of mathematics as well.

Some of Harriot's peers considered him a true universal genius whose broad-ranging work anticipated that of several more famous individuals. After his death, his editor said that if Harriot had "published all he knew of algebra, he would have left little of the chief mysteries of that art unhandled." Little is known of Harriot's personal life, although in a letter he wrote to German astronomer and mathematician **Johannes Kepler** (1571–1630; see entry in volume 2) he complained of ill health. His health was certainly not helped by his brief imprisonment in the Gatehouse of London, where he was held for suspicion of his being involved with a group of men who wanted to blow up the Houses of Parliament, or by the continual harassment he suffered when accused of using magical powers and of being an atheist (one who does not believe in the existence of God). By 1613, Harriot began to suffer from an ulcer in his nose, and was forced to give up his scientific work in 1614. This growth eventually proved to be cancerous and led to his death in 1621. According to his inscription on his tomb, Harriot "cultivated all the sciences / And excelled in all."

For More Information

Fox, Robert, ed. *Thomas Harriot: An Elizabethan Man of Science.* Burlington, VT: Ashgate, 2000.

Thomas Harriot

Franceschetti, Donald R., ed. *Biographical Encyclopedia of Mathematicians.* New York: Marshall Cavendish, 1999.

Lohne, J. A. "Thomas Harriot." *Biographical Dictionary of Mathematicians.* New York: Charles Scribner's Sons, 1991, pp. 985–90.

Mathematicians and Computer Wizards. Detroit: Macmillan Reference USA, 2001.

O'Connor, J. J., and E. F. Robertson. "Thomas Harriot." *The MacTutor History of Mathematics Archive.* School of Mathematics and Statistics, University of St. Andrews, Scotland. http://www.groups.dcs.st-andrews.ac.uk/~history/Mathmaticians/Harriot.html (accessed June 26, 2002).

Shirley, John William. *Thomas Harriot, a Biography.* New York: Oxford University Press, 1983.

Staiger, Ralph C. *Thomas Harriot, Science Pioneer.* New York: Clarion Books, 1998.

Born 470 B.C.E.
Chios (now Khíos), Greece

Died 410 B.C.E.
Larissa, Thessaly, Greece

Greek geometer

Hippocrates of Chios

Hippocrates of Chios is called the greatest mathematician of the fifth century B.C.E. He wrote what scholars consider the first mathematical textbook, studied and solved some of the classic problems of mathematics, and was among the first to show how logical proofs are developed in **geometry** (see entry in volume 1).

From merchant to mathematician

Hippocrates (pronounced hih-POCK-ruh-teez) of Chios (pronounced KY-oss or KEE-oss) gets his name from the small Greek island of Chios (now Khíos) where historians believe he was born. He is often confused with the better-known Greek physician Hippocrates (460–370 B.C.E.), who lived about the same time and is called the father of medicine. To make matters more confusing, Hippocrates the physician was from the Greek island named Cos, and is often referred to as Hippocrates of Cos. Nothing is known about the youth and early years of Hippocrates of Chios, and what scholars do know of his life is the result of what others have written about him.

As told by a sixth-century Greek scholar, Hippocrates of Chios was a successful merchant who lost all of his property after pirates captured him. Others say that he lost his goods when he was attacked by Athenian (pronounced uh-THEE-nee-an) pirates near Byzantium (pronounced bih-ZAN-tee-um), an ancient city located in the present-day Turkey province of Istanbul. Yet another story says that Hippocrates was swindled (tricked) out of his money by dishonest customs officials. Whatever the reason, all agree that his financial misfortune led him to go to the Greek city of Athens where he tried to get his money back through the courts. As this took a very long time, he chose to remain in Athens, which was then a great center of learning. Among those living and teaching in Athens at this time were Greek philosopher Socrates (pronounced SOK-ruh-teez) (470–399 B.C.E.) and his devoted pupil, Plato (427–347 B.C.E.).

Although most historians think that it was in Athens that Hippocrates of Chios received his real education in mathematics by attending lectures, others feel that he had already learned his mathematics in Chios. This is so, they argue, because it is known that he was greatly influenced by the Pythagorean (pronounced puh-thag-uh-REE-un) school of mathematics, and Samos, which is very near Chios, was the birthplace of the famed Greek mathematician, Pythagoras (560–480 B.C.E.). However, if Hippocrates had been trained as a Pythagorean, he would never have considered earning money from his own teaching, since Pythagoreans believed that taking money for giving out knowledge was forbidden. On the contrary, Hippocrates of Chios founded a school in Athens where he became one of the first to support himself openly and publicly by taking money for teaching mathematics.

Writes first mathematics textbook

Hippocrates of Chios was a pioneer in other ways as well, having written what is regarded as the first mathematical textbook. Although none of his written work survives, it is known that he wrote a book called *Elements of Geometry.* By the middle of the fifth century, when he was teaching, Greek mathematics had expanded so fast that there was little or no way to study it all in a logical manner. Hippocrates decided to organize all of geometry into one book, and his pioneering work had a considerable influence on all who followed. Not only did his book provide a model for the classic work of the same title by Greek mathematician

Euclid of Alexandria.
*Painting by Justus van Ghent.
Reproduced by permission of
Corbis-Bettmann.*

Euclid of Alexandria (325–270 B.C.E.; see entry in volume 1), but
it was the first to develop geometrical theorems (formulas) in a
logical manner from axioms and postulates (facts widely accepted
as true). This means that he began the now-familiar pattern of presenting geometry as a chain of propositions (statements). This
method showed how propositions could be derived on the basis of
an earlier one that has already been proven. His organizing textbook gave mathematicians some form of structure, since they now
knew which theorems depended on which of the others. He called

his book *Elements* because he had put together the "elements" of geometry, or those basic principles upon which all the other theorems were based. Although his book has not survived, historians know about it because it was referred to by the many mathematicians who followed him and who used it often. Most now agree that Euclid's *Elements,* which went on to become the most famous mathematical textbook of all time, was based on the work of Hippocrates of Chios.

Classic problems of mathematics

In some ways, mathematics is different from other scientific disciplines in that it has traditionally challenged itself to solve certain very difficult problems. Sometimes these problems are so difficult or complex that the people who pose them are not even sure if a solution is possible. However, whether or not these problems are solved or any solutions are ever found, this tradition guarantees that the best mathematical minds have something substantial to stimulate them, and that a new generation can be connected to the older ones by working on the same "classic" problems. During the time that Hippocrates of Chios was in Athens, one of the problems that fascinated mathematicians was the ancient one called "squaring the circle." The question and the challenge had already stood unsolved for a long time: given a **circle** (see entry in volume 1), and using only a straightedge and a compass, construct a square that has an **area** (see entry in volume 1) equal to that of the circle.

In attempting to solve this problem, Hippocrates of Chios introduced to geometry a very useful technique now called the method of exhaustion. Basically, this method translates a difficult problem into a simpler version, solves this simpler form, and then applies the solution to the more difficult problem. In trying to square the circle, Hippocrates first solved how to square a "lune," which is a crescent-shaped arc or a part of a circle. After achieving this with some difficulty, he then argued that the same method could therefore be applied to a circle, and that he had solved this classic problem. However, later generations realized that he did not in fact solve the problem since they were able to show that his solution did not apply to every type of lune, but only to those of a certain shape. Thereafter, "squaring the circle" became a challenge for one generation of mathematicians after another, until the late nineteenth century when German geometer Carl Louis Ferdinand von Lindemann (1852–1939) was able to prove without a doubt that squaring the circle was impossible.

Despite the fact that he did not solve this classic mathematical dilemma, the contributions of Hippocrates of Chios to mathematics were considerable and significant. Many historians also attribute to him the very useful indirect method of proof, in which one first assumes that the opposite of what is wanted to be proven is false. Therefore, by proving the opposite to be false, the alternative is then considered to be true. Hippocrates of Chios also studied astronomy as he grew older. He stands as an excellent example of the notion that it is never too late to learn new things and to push the mind's boundaries.

For More Information

Bulmer-Thomas, Ivor. "Hippocrates of Chios." In *Biographical Dictionary of Mathematicians.* New York: Charles Scribner's Sons, 1991, pp. 1083–91.

Burton, David M. *The History of Mathematics: An Introduction.* New York: McGraw-Hill, 1999, pp. 115–20.

O'Connor, J. J., and E. F. Robertson. "Hippocrates of Chios." *The MacTutor History of Mathematics Archive.* School of Mathematics and Statistics, University of St. Andrews, Scotland. http://www.groups.dcs.st-andrews.ac.uk/~history/ Mathematicians/Hippocrates.html (accessed June 26, 2002).

Born December 10, 1804
Potsdam, Prussia (now Germany)

Died February 18, 1851
Berlin, Germany

German mathematical physicist

Carl Jacobi

Carl Jacobi.
Reproduced by permission of the
Granger Collection Ltd.

Carl Jacobi founded the theory of elliptic functions and carried out important research on number theory and differential equations. He was a gifted and highly popular teacher who is credited with developing the modern method of teaching by seminar. His mathematical advances had an immediate effect on physics.

Home-schooled prodigy

Carl Gustav Jacob Jacobi (pronounced yaw-KOH-bee) was born in Potsdam, Prussia, which is now part of Germany. He came from a very wealthy family, and his father, Simon Jacobi, was a prominent banker. All that is known of his mother, however, is that her maiden name was Lehmann. Jacobi had two brothers, Moritz and Eduard, and one sister, Therese. Young Jacobi was a prodigy (pronounced PRAH-dih-gee; a highly talented child), and his early education took place at home where he was tutored by an uncle. As an eleven-year-old, he entered a gymnasium (secondary school that prepares pupils for college) in Potsdam, Prussia. Although this is usually a nine-year school, which students leave at the age of nineteen or twenty, Jacobi was found to be so advanced that he was placed in the final (highest) class. At the end of that academic year, Jacobi graduated at the age of twelve. Although he was thought to be ready to enter the University of Berlin, that school's policy would not allow it to accept students under the age of sixteen. This meant that Jacobi had to remain in the same class at the gymnasium until the spring of 1821.

His time at the gymnasium was not wasted, however, as he spent these years reading very advanced mathematical texts. He also received awards in Latin, Greek, and history during this time. By the time he was ready to enter the University of Berlin, he had read and mastered a great deal of the advanced mathematics of his time. One result of this was that he rebelled against his teachers at school and continued to read what he thought was most important. He also conducted research on his own since he felt the level of mathematics at Berlin was not very high. Within a year, Jacobi had graduated with top marks in classical languages and history, as well as in mathematics. By 1823, he had passed the examinations necessary for him to be able to teach Greek, Latin, and mathematics in a gymnasium, and he was at the same time already finishing work on his doctoral thesis.

Begins teaching career

In 1825, Jacobi earned his doctorate from Berlin and accepted a teaching position at one of the leading gymnasiums in Berlin. This was unusual, not because of his age, but because Jacobi was Jewish, and Jews were rarely allowed to teach at gymnasiums. It appears that he was given the position simply because of his incredible ability and intelligence. That same year, however, Jacobi decided to give up his Jewish faith and convert to Christianity. This made him eligible to teach at the university level as well, and he immediately obtained a teaching job at the University of Berlin for the academic year 1825–26. He did not stay in Berlin very long and after taking the advice of colleagues and friends, he accepted a position as lecturer at the University of Königsberg in May 1826. (Königsberg is on the Baltic coast of Germany and is now called Kaliningrad and is part of Russia.) By this time, Jacobi had been corresponding with some of the top mathematicians in Europe for quite a while, and it was they who offered advice and recommendations when he needed it.

At Königsberg, he became assistant professor in 1827 and was made full professor in 1832. Jacobi came into his own as a

Carl Jacobi

Carl Jacobi

teacher, and he proved to be a truly gifted and enthusiastic professor. In 1831, Jacobi married Marie Schwinck, with whom he would have seven children. At Königsberg, his reputation as an excellent teacher attracted students from far and wide, and it was there that he developed what is now recognized as the modern method of teaching by seminar. A seminar is when a professor meets regularly with a group of advanced students to discuss their original research. Another method that the brilliant Jacobi used was to present himself to his students as one who knows little and who desires to know more. He found that this technique seemed to inspire his students to be like him and to not be satisfied with what they knew, but instead to always want to learn more. He also encouraged them not to be afraid to attempt their own independent research in mathematics too early, before they had mastered everything, telling them that just as they do not have to meet all the girls in the world before marrying one, they do not have to "meet" all the subjects in mathematics before "marrying" one.

Mathematical contributions

Jacobi's contributions range across several branches of mathematics, but it was with his first major publication in 1829, *Fundamenta nova theorae functionum ellipticarum* (New Foundations of the Theory of Elliptic Functions), that he really made his mark. At the time, Norwegian algebraist **Niels Abel** (1802–1829; see entry in volume 1) was doing the same sort of work on elliptic functions, but neither was aware of the other's work. Elliptic functions are not ellipses, which are oval shapes, but instead are rules used to calculate something. Originally, they were used to calculate planetary orbits, but they would eventually prove very useful to other branches of mathematics as well.

There was hardly an important branch of mathematics that Jacobi did not touch upon, and overall, his mathematics had an immediate effect on classical mechanics, as originally put forth by English mathematician and physicist **Isaac Newton** (1643–1727; see entry in volume 2), and as continued and elaborated upon by French algebraist and number theorist **Joseph-Louis Lagrange** (1736–1813; see entry in volume 2) and French mathematical physicist, statistician, and astronomer Pierre Simon Laplace (1749–1827). Classical mechanics is a part of pre-atomic or older physics that studied the forces and motions of familiar bodies and things that are known and can be seen in the everyday world. Classical mechanics is usually contrasted with quantum or atomic physics, which deals with the motion of things that cannot be seen, such as subatomic particles like neutrons, protons, and electrons. Still later, in more modern times, Jacobi's mathematical work would have an effect on the quantum mechanics and relativity theories of the twentieth century.

Financial and political difficulties

Jacobi's father died in 1832 and left a small fortune for his family to inherit. The family lived well for another eight years until 1840 when a severe economic depression in Europe led to the family's bankruptcy. Jacobi was always one to work too hard, and by 1842 he suffered a collapse, after which it was discovered that he had diabetes. In addition to his worsening health, Jacobi found that for the first time in his life he was dependent on his teaching salary. Fortunately, his mathematical colleagues around Europe came to his aid and persuaded the king of Prussia, Frederick William IV (1795–1861), to give Jacobi a royal pension. This would not only help him provide for his mother, wife, and seven children, but would allow him to go to Italy for his health as his doctors recommended. The king also allowed him to remain in Berlin rather than return to Königsberg where the weather was harsher.

As his health improved, Jacobi took an interest in the revolutionary movements that were sweeping across Europe and in 1848 he entered politics by running for local office in Berlin. However, he ran as a member of the liberal party, making him suspect to both fellow liberals and his opponents. He offended the king's supporters because he ran as a republican or a liberal. Other liberals were offended and distrustful of him because he was living off a royal pension. Jacobi lost the election and his pension was temporarily suspended. It was eventually returned, however, to prevent the University of Vienna from hiring him away. By 1850, Jacobi's health was again declining, and in 1851, at the age of forty-six, he first contracted influenza (the flu), followed by a serious case of smallpox. Already suffering from diabetes, Jacobi died in Berlin that same year. In 1852, he was memorialized by his many friends and colleagues as one of the greatest members of the Berlin Academy of Sciences.

Carl Jacobi

Carl Jacobi

For More Information

Franceschetti, Donald R., ed. *Biographical Encyclopedia of Mathematicians.* New York: Marshall Cavendish, 1999.

O'Connor, J. J., and E. F. Robertson. "Carl Gustav Jacob Jacobi." *The MacTutor History of Mathematics Archive.* School of Mathematics and Statistics, University of St. Andrews, Scotland. http://www.groups.dcs.st-andrews.ac.uk/~history/Mathematicians/Jacobi.html (accessed June 26, 2002).

Born February 22, 1928
Oak Park, Illinois

American statistician and computer scientist

Thomas E. Kurtz

Thomas Kurtz designed and developed, along with American mathematician John Kemeny (1926–), the **computer** (see entry in volume 1) programming language known as BASIC. This user-friendly programming language became the first general-purpose language that beginners could learn and start to use almost immediately. BASIC became the most widely used programming language in the world.

Thomas E. Kurtz.
Reproduced by permission of
Thomas E. Kurtz.

Majors in mathematics

Thomas Eugene Kurtz was born on February 22, 1928, in Oak Park, Illinois. His father, Oscar Christ Kurtz, worked in various capacities at the International Lion's Club headquarters, and his mother was Helen Bell. As a youngster, Kurtz was interested in science, and when he entered Knox College in Galesburg, Illinois, he intended to major in physics. Since he also intended to take all of the mathematics courses available at Knox, an adviser suggested that he might consider a career in **statistics** (see entry in volume 2). This, he said, would allow Kurtz the opportunity to apply his mathematical skills to a variety of scientific problems. Following this advice, Kurtz switched his major from physics to mathematics in his senior year, and graduated from Knox College in 1950 with a bachelor's degree in mathematics.

Thomas E. Kurtz

Early computer experience at Princeton

Kurtz went directly from Knox College to Princeton University in New Jersey for his graduate education. At Princeton, his interest in computing was stimulated by a professor of engineering named Forman Acton. The year 1951 marked the very early beginnings of computer progress, and the first fully electronic computers were just starting to begin operations. It was through Acton that Kurtz was able to spend a summer at the Institute of Numerical Analysis, a branch of the National Bureau of Standards located on the campus of the University of California at Los Angeles (UCLA). This was a unique opportunity for a young man interested in computers, as Kurtz not only attended lectures on computing but also was able to meet and work with some of the early computer pioneers who were at UCLA that summer. While studying for his doctorate at Princeton, Kurtz worked from 1952 to 1956 as a research assistant in the Analytical Research Group on campus where he wrote computer programs to help solve classified military research problems. Kurtz ran these programs on an IBM Card Programmed Calculator, and occasionally his job involved tending this huge machine throughout the night, transferring cards from the output bin back to the input holder.

Joins Kemeny at Dartmouth

When Kurtz graduated from Princeton in 1956 with a Ph.D. in mathematics, he was recruited by John George Kemeny, the chair of the mathematics department at Dartmouth University in Hanover, New Hampshire. The Hungarian mathematician had emigrated to the United States in 1940 and worked on the Manhattan Project in Los Alamos, New Mexico, to build an atomic bomb. At the end of World War II (1939–45), Kemeny went to Princeton for his doctorate and worked there with German American physicist **Albert Einstein** (1879–1955; see entry in volume 1). It was at Princeton that Kemeny became convinced that computers would eventually become a key tool in education. One of Kurtz's first assignments at Dartmouth was as the contact person between Dartmouth and the New England Regional Computer

Center, which had been established at Massachusetts Institute of Technology (MIT). International Business Machines (IBM) had funded the center, which would provide local colleges with access to its facilities. Kurtz took advantage of this opportunity and spent August 1956 at MIT learning assembly language programming. This is the highly technical language that the center's computer, an IBM 704, understood. At this point in time, the official computer languages that existed were designed for professionals and took months to learn.

In 1959, Dartmouth College purchased a computer of its own and Kurtz was appointed director of computing. At first, the new computer was used by only a small fraction of the students and faculty, but Kurtz felt that everyone at Dartmouth should be able to use its computing facilities. He noted, however, that one of the drawbacks to this idea was that users could not hold or reserve time on a given machine, but instead had to submit their programs to be processed by someone else at the computer center. This system operated by running each program in the order it was received. It would then store the results. This process was called "batch processing," and meant that users had to wait as much as a day or more just to see their results. It also meant that any corrections or "debugging" that they had to make in their program had to be resubmitted, and therefore took even more time.

Time-sharing idea

As Kurtz sought a way around this roadblock, he brought several somewhat radical ideas to the new notion of computing. One of the major debates he had with his Dartmouth colleagues was about how to charge students for computer time. On this subject, Kurtz had a radical answer, which was ahead of its time. In his 1985 book, *Back to BASIC*, he wrote, "I would be almost the only one arguing that computing costs should be part of tuition, just like library costs. My point was that computing was a fixed cost— it cost the same whether it was used or not—so why not let students use it?" Kurtz put this and other ideas to Kemeny, and received a welcome response, especially when Kurtz explained his concept of time sharing. Time sharing allowed many people to use a computer at the same time by having the computer work on each person's problem for short periods of time. Although the central computer is actually dividing its time between users, it gives people the illusion that they have private access. Time sharing had been used on large projects at big companies, but no general-purpose time sharing system was available for the varied computing needs of a place like Dartmouth.

In 1963, Kemeny and Kurtz began the planning stages for a new campus computer system, and in February 1964 they decided to develop a time sharing computer system with the General Electric Corporation. By June of that year, they had completed the Dartmouth Time Sharing System, which would handle communications with terminals all over campus. Their goal was to make access to computing as simple as checking out a book in the college library. The system they designed gave all Dartmouth students, and even students from other area schools, access to the central computer whenever they wished, without the obstacles of forms, permission, and restricted hours.

Designs BASIC language

Once Kurtz and Kemeny had removed most of the barriers to computer use, they realized that the existing computer languages, like

Thomas E. Kurtz

FORTRAN (Formula Translation) and ALGOL (Algorithmic Language), had been created for engineers and scientists and were very complicated and difficult to learn. They knew from experience that they were far too complex for the average person to learn without months of training. What was needed, they realized, was a completely new language that was simple enough for beginners to learn quickly, yet flexible enough to handle all different kinds of applications. Kurtz and Kemeny set out to create a general-purpose language that beginners could learn and start using almost immediately. Their goal was to make computer programming accessible to nearly anyone.

In developing a new language, Kurtz and Kemeny naturally started with elements from the two existing languages, FORTRAN and ALGOL, but soon added many commonsense, original features such as line numbers that made it much easier to pinpoint and correct errors. With the goal of simplicity always in mind, their new language had only fourteen commands, and when they tried it out on students, they found that they could begin programming after only two lessons. On May 1, 1964, the Dartmouth Time Sharing System, with its BASIC (Beginners All-Purpose Symbolic Instruction Code) programming language, became operational. By June 1968, more than eighty percent of the undergraduates at Dartmouth could write BASIC programs. Since it was not written in machine language, BASIC soon became hugely popular and eventually became the most widely used language in the world.

Since neither Kurtz nor Kemeny took out a patent or a copyright for their new language, others soon developed many different versions of BASIC. For example, in 1975, Microsoft was founded by Bill Gates (1955–) and Paul Allen (1953–) to market a program they had written that allowed the Altair 8800 personal computer to understand BASIC. In 1983, Kurtz and Kemeny formed True BASIC, Inc., with the intention of creating a much-expanded personal computer version of BASIC for educational purposes. Although still used in some high schools and elementary schools, this version of BASIC has gradually declined over the years in favor of other languages.

Post-BASIC career

While serving as professor of mathematics and computing at Dartmouth, Kurtz served as director of the Kiewit Computation Cen-

Thomas E. Kurtz

Thomas E. Kurtz

ter at Dartmouth from 1966 to 1975, director of the Office of Academic Computing from 1975 to 1978, and vice-chair (1979–83), and chair (1983–88), of the Program in Computer and Information Science. Over the years, Kurtz served as principal investigator for various projects supported by the National Science Foundation to promote the use of computers in education. He also participated in many other activities related to the use of computing in teaching, including the Pierce Panel of the President's Scientific Advisory Committee. In 1991, he was recognized as a computer pioneer by the Institute of Electrical and Electronics Engineers (IEEE), and was made a fellow of the Association for Computer Machinery (ACM) in 1994. Kurtz remains professor emeritus at Dartmouth University and concentrates his research on statistics.

For More Information

"Back to BASIC." *Technology Review.* Massachusetts Institute of Technology. October 2001. http://www.techreview.com/articles/trailing1001.asp (accessed June 27, 2002).

Kemeny, John G., and Thomas E. Kurtz. *Back to BASIC: The History, Corruption, and Future of the Language.* Reading, MA: Addison-Wesley, 1985.

"Thomas E. Kurtz." *Jones Telecommunications & Multimedia Encyclopedia.* http://www.digitalcentury.com/encyclo/update/kurtz.html (accessed June 27, 2002).

Travers, Bridget, ed. *World of Invention.* Detroit: Gale, 1994, p. 352.

Born August 25, 1728
Mulhausen, Alsace (now part of France)

Died September 25, 1777
Berlin, Prussia (now Germany)

Swiss-born German statistician, geometer, and analyst

Johann Lambert

Johann Lambert is best known in mathematics for being the first to provide a rigorous proof that **pi** (see entry in volume 2) is an **irrational number** (see entry on rational and irrational numbers in volume 2). Mostly self-taught, he was a man of great ability and a many-sided scholar who did very high-level work in several scientific disciplines. He was the first to devise accurate methods for measuring light intensities.

Humble beginnings

Johann Heinrich Lambert (pronounced LAHM-burt) was born in the town of Mulhausen, now a part of northeast France, but then a free city allied with Switzerland. His background was very humble since both his father and grandfather were poor tailors. His father, Lukas Lambert, and his mother, Elizabeth Schmerber, had four other boys and two girls. Since the family was so poor, Lambert's father took his son out of school when he was twelve years old so that he could earn some money for the family. Although young Lambert had received only a few years of elementary instruction, he learned enough French and Latin to be able to continue to study without a teacher. This was something he would do

Johann Lambert

throughout his life, as he would study and read on his own late at night or during any spare time he had during work.

At the age of fifteen, his excellent handwriting earned him a job as a clerk at a local ironworks, and two years later, he became secretary to the editor of a newspaper printed in Basel, Switzerland. In 1748, a year after Lambert's father died, he was appointed tutor to a private family on the recommendation of the newspaper editor. Lambert then moved to Chur (pronounced KUR) in eastern Switzerland where the family of Peter von Salis lived. Von Salis had been ambassador to the English court and was married to an Englishwoman. Lambert tutored three children ranging in ages from seven to eleven, and would remain their tutor for the next ten years. This period would prove to be a very important time for Lambert as his job allowed him enough time off to be able to take advantage of the family's excellent library. As an independent learner, he was able to study and read intensively during these years. Since he was employed by a well connected family, he was also able to meet many of the family's friends and visitors, some of whom were very learned individuals.

An educational journey

In 1756, Lambert took two of his oldest pupils in the von Salis family on a *Bildungsreise* (educational journey) through parts of Europe. This was considered a necessary part of a proper young person's education. For the next two years, Lambert and his pupils visited many of the major cities of Germany, Holland, France, and Italy. During this time, Lambert made his own personal *Bildungsreise* as well. For example, in Germany, he attended lectures and meetings, and studied the works of some of the best scientists. In Holland, he met and visited some of the top physicists of the time. By 1758, his pupils were grown and Lambert was ready to find a more permanent job. What he had in mind was a scientific position teaching at the University of Göttingen in Germany, but this did not happen. All his life, Lambert would be considered somewhat of an eccentric or a peculiar person who dressed strangely and did not always follow the conventions or ways of everyday society. This did not help him in Germany, and when he later tried to settle in Zurich, Switzerland, he was also considered to be too much of a non-conformist.

Joins Prussian Academy

For the next five years or so, Lambert did not live a settled life, as he traveled from Germany to Switzerland in search of an appropriate position. During his earlier trips, Lambert always made contact with the various scientific societies in the towns and cities he visited, and when a new Bavarian Academy of Sciences was being planned, Lambert was selected to help organize the academy in Munich. When he left this position in 1762 after having differences with the members, he took a position as a geometer in a Swiss survey to establish the frontier between the Italian city of Milan and the Swiss city of Chur, all the while hoping to secure a permanent position with the older and more established Prussian Academy of Sciences. He finally obtained this position in 1765, and eventually became a favorite of the king of Prussia, Frederick the Great (1712–1786). It was under the support of Frederick that Lambert would write more than 150 papers and do his best and most productive work.

Mathematical contributions

Lambert developed many basic concepts in mathematics, but he is best known for being the first to provide a precise proof showing that pi is an irrational number. Pi is the constant **ratio** (see entry in volume 1) of the circumference of a **circle** (see entry in volume 1) to its diameter. A rational number is one that can be expressed as the quotient of two **whole numbers** (see entry in volume 2). An irrational number, therefore, cannot be expressed as a ratio and when carried out, it simply goes on forever. In 1767, Lambert was able to prove and demonstrate precisely that no **fraction** (see entry in volume 1) or **decimal** (see entry in volume 1) number can *exactly* represent pi.

In **geometry** (see entry in volume 1), Lambert seemed to have great intuitive (ability to understand something immediately and directly) powers. When he tried to prove the parallel postulate (something that is given or widely accepted) of Greek geometer and logician **Euclid of Alexandria** (c. 325–c. 270 B.C.E.; see entry in volume 1), he did work that foreshadowed the discovery some seventy-five years later of what came to be called non-Euclidean geometry. Lambert also wrote a book for artists in 1759 that is considered to be a masterpiece in the field of descriptive geometry.

Johann Lambert

Johann Lambert

Descriptive geometry is sometimes used interchangeably for projective geometry, as both involve the representation of three-dimensional objects in two dimensions. However, descriptive geometry formed the basis for all engineering and architectural drawings, while projective geometry deals literally with things that are projected, such as motion pictures, maps, and shadows.

Lambert also made major contributions to the theory of map construction with his work on map projections. In order to make a useful and realistic map of the Earth, mapmakers must "project" geometric figures from a sphere (since the real world is round and three-dimensional) onto a flat sheet of paper. This involved some of the same problems of perspective that an artist used to face in trying to make a painting look more real. In this field, Lambert was the first to state the mathematical conditions for map projections. A lesser-known mathematical contribution was his work in producing a general theory of errors. His work experimenting with how often errors will be made in different systems would form the basis for the work of statisticians more than a century later.

Other achievements

During his lifetime, Lambert wrote so much about so many different areas of science and mathematics that much of his work was not fully known until it was published long after his death. Apart from his mathematical contributions, one of the discoveries for which he will be forever recognized is in his study of light. In *Photometria* (Photometry), his famous book on the measurement of the intensity of light, he laid the foundations for the modern study of optics (the study of light and its effects). It was in this work that he became the first to discover techniques for accurately measuring light intensity. In fact, the metric unit of brightness is named after him. In 1761, he published an article on comets. In it, he speculated that the stars in the neighborhood of the Sun made up a connected system, and that it was groups of such systems that made up what is now called the Milky Way galaxy. He also thought that there could be other such groupings far out into space. Years later, his guesses would prove to be correct.

Toward the end of his life, Lambert conducted meteorological (weather-related) studies in which he tried to find a way to reliably measure humidity. He did the same with the radiation and reflec-

tion of heat. Lambert was also a philosopher who wrote on the theory of knowledge. In this field, he corresponded with German philosopher Immanuel Kant (1724–1804), who many consider one of the most influential thinker's of modern times. Overall, the self-taught Lambert demonstrated remarkable ability and achievement in every subject or discipline he studied. It is said that when Frederick the Great asked him which science he was best at, Lambert replied simply, "All."

A spherical halo of uncharged particles surrounds the Milky Way galaxy. Johann Lambert studied the Milky Way and comets. *Courtesy of the National Aeronautics and Space Administration.*

For More Information

O'Connor, J. J., and E. F. Robertson. "Johann Heinrich Lambert." *The MacTutor History of Mathematics Archive.* School of Mathematics and Statistics, University of St. Andrews, Scotland. http://www.groups.dcs.st-andrews.ac.uk/~history/Mathematicians/Lambert.html (accessed June 26, 2002).

Johann Lambert

Scriba, Christoph J. "Johann Heinrich Lambert." In *Biographical Dictionary of Mathematicians.* New York: Charles Scribner's Sons, 1991, pp. 1323–28.

Wilkins, D. R. "Johann Heinrich Lambert." *The History of Mathematics.* From *A Short Account of the History of Mathematics.* W. W. Rouse Ball. http://www.maths.tcd.ie/pub/HistMath/People/Lambert/RouseBall/RB_Lambert.html (accessed June 27, 2002).

Born December 10, 1815
London, England

Died November 27, 1852
London, England

English applied mathematician

Ada Lovelace

A lthough born in the early 1800s, Ada Lovelace is today credited with writing the world's first **computer** (see entry in volume 1) program. Having met English mathematician and inventor **Charles Babbage** (1791–1871; see entry in volume 1) when
she was only eighteen, she immediately understood how his proposed computing machines (which he called "engines") would
work and wrote an article in which she outlined the basic concepts of computer programming. She also described the main elements needed for any computer language and even made predictions about how computers would be used. Although she wrote
far in advance of Babbage or anyone else actually building something like today's all-purpose computers, she understood so well
how computers would work in principle and was able to explain
this so clearly, that when her work was rediscovered in the 1950s,
it became the final word in explaining the process now known as
computer programming.

Ada Lovelace.
Public domain.

Unusual home life

Nothing about Augusta Ada Lovelace's life was ordinary. Her father
was English poet George Gordon, Lord Byron (1788–1824), and

Ada Lovelace

her mother was a wealthy aristocrat named Anne Isabella Milbanke. Known by her middle name, Ada's parents separated for good only one month after she was born, and she never again saw her father. He left England and died in Greece of fever when she was eight years old. Her mother feared that young Ada might inherit her absent father's "poetic" personality—she believed he was not very dependable and too emotional—so she decided to have her trained in the sciences. Ada's mother especially favored mathematics, since she herself had been tutored in it and thought it would help Ada discipline her mind. When Ada showed she was capable of spelling and adding six rows of numbers at the age of five, her mother arranged some of the best mathematicians in England to tutor her.

Although Milbanke was an intelligent woman and held very strong opinions about certain issues, she sometimes did odd things while raising Ada. One of these was her insistence that her young daughter lie perfectly still on a wooden plank for long periods of time without even moving a finger in the belief that this taught self-control. Despite this, Ada was a very active young girl and loved gymnastics as well as dancing and riding horses. She also was fluent in French, could draw very well, and could play the piano, the violin, and the harp. As a teenager, she attended the theater, concerts, and fashionable parties. When Ada was fifteen, her mother arranged for Scottish mathematician and astronomer Mary Fairfax Somerville (1780–1872) to tutor her. As noted in *The Bride of Science,* Somerville was known as the "queen of nineteenth-century science," and she became one of the first two woman elected to the Royal Astronomical Society in England. Somerville would greatly influence young Ada, since she and her two daughters would take Ada with them when they attended geography lectures at the University of London.

Lovelace meets George Babbage

When Lovelace was only seventeen, she became a member of a group of women who called themselves the Bluestockings. Much more than a social group, these women would attend serious lectures and discussions, and would often visit museums or the homes of famous scientists. It was at one of these events that Lovelace met Babbage, a brilliant if sometimes strange inventor who had thought out many of the basic principles of modern

Ada Lovelace

Mary Fairfax Somerville. The Scottish mathematician tutored the fifteen-year-old Ada Lovelace. *Reproduced by permission of Archive Photos, Inc.*

computers long before they could ever be built. Babbage demonstrated a small working model of his steam-driven calculating machine, and he also discussed plans for a much more ambitious machine he called his "analytical engine." According to Babbage, this machine would get its instructions and numerical data from punched cards, and it would be able to make and even analyze mathematical calculations. Lovelace was fascinated by the machines that Babbage had built, and actually understood how his planned machines would work. Babbage was immediately impressed with Lovelace's quick grasp of the basic concepts of his engines, and the two became good friends. Babbage arranged for Lovelace to receive special instructions in mathematics that were equal to what the bright young men of her time were learning at Cambridge University.

Writes "First" Computer Program

In 1842, when Lovelace was twenty-seven, Babbage traveled to Italy where he gave a lecture that so impressed a young Italian military engineer that the engineer wrote an article describing Babbage's ideas. Since the article was to be published in a Swiss journal, however, it was written in French. This article was important because Babbage had been too busy to write about his own accomplishments and plans, making this the first published summary of Babbage's groundbreaking work. Babbage's friends asked Lovelace, who knew both the subject and the language, to translate the article into English. When she discovered that the article only briefly described the mathematical concepts by which Babbage's machines would work, she decided to add a series of notes to the translation. Babbage encouraged her to do so, and when she finished, her "notes" turned out to be three times as long as the original article. Her addition would eventually become more important than the original article itself, since Lovelace produced not only the first clear mechanical explanation of Babbage's planned analytical engine, but also provided actual examples of how this machine might be instructed to perform certain tasks. By doing this, she created what is now recognized as the world's first computer program or set of instructions.

In July 1843, Ada's translation and notes were published in the English journal *Taylor's Scientific Memoirs.* Titled "Sketch of the Analytical Engine Invented by Charles Babbage, Esq.," it was signed only by the initials A.A.L., for Augusta Ada Lovelace. Although at the time few people knew that she had made an important contribution to mathematics with this article, the rediscovery of her work in 1953 indicated just how well she grasped and could explain the concept of computer programming. Her notes discuss many parts of a computer program that are today taken for granted. For example, she invented the idea of a "subroutine," a set of instructions that are used repeatedly in a variety

Ada Lovelace

English mathematician Charles Babbage, with whom Ada Lovelace worked closely. *Reproduced by permission of the Corbis Corporation.*

of ways. She also anticipated the process she called "backing" which is known as "looping" today. A loop is a technique that allows a program to repeat a series of instructions a number of times. Ada also came up with the concept of "garbage in, garbage out," since she warned that a computer could do nothing to correct the fact that the user had entered untrue information. She also described the notion of "conditional jump," which is the same as today's "if-then" rules of **logic** (see entry in volume 2) that a computer program would follow. Besides these and several other concepts, she also offered several predictions, stating that engines like Babbage's would eventually be used to compose complex music and to produce graphic images, and that they would have as many scientific uses as they do practical uses.

Personal problems and difficulties

When Lovelace signed her 1843 article A.A.L., she did so because she had been married to William King, first earl of Lovelace since 1835. When her husband joined the House of Lords in 1838 and took that title, she became the countess of Lovelace. The couple had three children between 1836 and 1839, and for a time lived a privileged life. However, Ada Lovelace contracted cholera (pronounced KAHL-err-uh), a bacterial disease that often kills through severe diarrhea and massive dehydration, and although she somehow survived, she was never the same. Suffering from asthma and digestive problems, her doctors gave her painkillers, several of which became addictive, like laudanum (pronounced LOW-duh-num), which was made from the highly addictive drug opium. Since she often took the painkillers with wine, she often suffered from hallucinations and bizarre mood swings whenever she stopped taking them. She later developed a compulsive gambling habit, believing that she had found a secret mathematical gambling system for betting on horse races. However her system did not keep her from losing much of the family fortune, and she was heavily in debt when she died of uterine cancer in 1852. She was buried next to her father, Lord Byron, who had written a poem about her before he died (one excerpt: "I see thee not. I hear thee not. But none can be so wrapt in thee"). She and her father shared not only the same dark, good looks and difficult personal life, but both died in their thirty-sixth year.

Honored after death

It took a very long time for the world to catch up with Ada Lovelace's advanced ideas about computers. In fact, she had no direct influence on the development of computers, since her work was unknown, and it was only a century after her death that her work was read and its significance understood. Since the computer pioneers of the twentieth century did their research ignorant of her writings, her work has mostly historical interest. Yet in 1980, her work finally received public recognition as the U.S. Department of Defense officially named a software language "Ada" in her honor. Three years later, the American National Standards Institute (ANSI) approved Ada as a national, all-purpose standard.

For More Information

Baum, Joan. *The Calculating Passion of Ada Byron*. Hamden, CT: Archon Books, 1986.

Moore, Doris Langley-Levy. *Ada, Countess of Lovelace: Byron's Legitimate Daughter*. New York: Harper & Row, 1977.

Stein, Dorothy. *Ada: A Life and a Legacy*. Cambridge, MA: MIT Press, 1985.

Toole, Betty. "Ada Byron, Lady Lovelace." *Biographies of Women Mathematicians*. Agnes Scott College. http://www.agnesscott.edu/lriddle/women/love.htm (accessed June 21, 2002).

Toole, Betty Alexandra, ed. *Ada, the Enchantress of Numbers: A Selection from the Letters of Lord Byron's Daughter and Her Description of the First Computer*. Mill Valley, CA: Strawberry Press, 1992.

Wade, Mary Dodson. *Ada Byron Lovelace: The Lady and the Computer*. New York: Dillon Press, 1994.

Woolley, Benjamin. *The Bride of Science: Romance, Reason, and Byron's Daughter*. New York: McGraw-Hill, 1999.

Born May 9, 1746
Beaune, France

Died July 28, 1818
Paris, France

French geometer and educator

Gaspard Monge

Gaspard Monge.
Courtesy of the Library of Congress.

One of the most famous mathematicians of his day, Gaspard Monge was also one of the most original of his age. His major mathematical achievement was his invention of descriptive **geometry** (see entry in volume 1) and its application to problems of construction. He also helped establish the metric system and was a founder of the École Polytechnique, the first modern engineering college.

Humble beginnings

Gaspard Monge (pronounced MOHNZH) was born into a humble family as his father, Jacques Monge, was variously described as a knife grinder, peddler, and merchant. His mother's name was Jeanne Rousseaux, and Gaspard was the oldest of her three sons. Monge's father believed strongly in the power of education, and he passed on that belief to his children. Young Monge was raised in the town of Beaune, which is part of the Burgundy region of France, and it was there that he and his brothers attended a school that was operated by a religious order called the Oratorians. This order was more liberal (open-minded) than most religious groups, and its school stressed mathematics and science as well as the humanities and religion. It was at this school that Monge first showed his brilliance, but his family already considered him a genius.

When he finished his schooling with the Oratorians in 1762, Monge attended the Collège de la Trinité in Lyons, France, where, despite being a seventeen-year-old student, he was made a physics instructor. After completing his education in 1764, he returned home to Beaune for the summer and while there, he volunteered to sketch an exact map of the town. The high quality of this detailed and accurate drawing so impressed a local military engineer that he recommended Monge for a place at the École Royale du Génie at Mézières, France. This school was a highly regarded military school and since Monge was a commoner (someone with no special rank), he could only be admitted as a draftsman or technician. This meant that he would never graduate as an officer.

Produces a work that changes his life

Monge began to attend the École Royale du Génie in 1765 as a draftsman, even though his beginning duties did not use any of his mathematical talents. Monge spent this time developing his own ideas of geometry, and when he was asked to use his talents about a year later, he quickly produced what was recognized as a revolutionary new system of drawing for engineers. Monge had been asked to draw up a plan for a fort whose gun placements could not be seen by the enemy and whose locations were the best for firing at invaders. Creating such plans always took a very long time and the work tedious and complicated.

To draw up his plan for a fort, Monge took an entirely different approach from the traditional arithmetic methods. He used a geometric method instead, devising his own graphic system for dealing with this project. What he cleverly realized was that by using geometry instead of arithmetic, he was able to develop a systematic means of representing and analyzing three-dimensional geometric objects (like a real fort) on a flat, two-dimensional surface (like any piece of drawing paper). What Monge had invented would become known as descriptive geometry and would eventually be used in all modern mechanical drawing. Descriptive geometry is sometimes used interchangeably for projective geometry, as both involve the representation of three-dimensional objects in two dimensions. However, descriptive geometry formed the basis for all engineering and architectural drawings, while projective geometry deals literally with things that are projected, such as motion pictures, maps, and shadows. Monge's invention would transform this humble draftsman into one of the most influential and well-known mathematicians of his time.

At the time of his invention, however, Monge produced a well-drawn graphical sketch of the fort and its gun placements so quickly that his superiors suspected some form of trickery must have been involved. No one had ever been able to produce such a complete drawing in so short a time. In fact, his commanding officer at first even refused to look at his work, assuming that since it took so little time it had to be rushed and careless. However, once Monge was able to convince his superiors that his drawings involved an entirely different technique of representing three-dimensional objects in two dimensions, they not only recognized the superiority of his method and value of his discovery, but immediately declared it a top military secret. He was then directed to teach his method only to French military engineers, and was not allowed to teach it publicly until 1794, after the French Revolution (1789–93; a period during which the monarchy was overthrown and a republic was established).

Applies descriptive geometry

Monge was soon recognized as a mathematician of exceptional ability, and in January 1769, he was appointed professor of mathematics at the Royal School of Engineering at Mézières. The following year, he was also given the position of professor of physics. Monge would remain at this school until 1783, and would leave behind a record as a first-class teacher who was responsible for molding a brilliant French engineering group. During these years, Monge was able to systematize the basic principles of descriptive geometry and to show how to apply it to all sorts of graphical problems. What came to be known as "Monge's procedure" was shown capable of solving any questions of size, shape, or orientation by using graphical methods. Eventually, this procedure would form the basis for the "orthogonal projections" of all engineering drawings as well as the basis for every engineering course of study until the invention of computer graphics in the 1960s.

In 1783, Monge was appointed examiner of naval cadets, and when he was not inspecting naval schools, he was at his post in Paris. By this time, Monge had become part of the French scientific establishment, especially since his election to the Académie des Sciences in 1780. In 1777, Monge had married Catherine Huart, with whom he would have three daughters. Between his 1783 appointment and the French Revolution, Monge became interested and wrote a good deal about many subjects that were not necessarily mathematical. For example, he wrote several papers on mechanics, technology, and chemistry, and also worked on caloric (heat) theory, optics (the science of light), and acoustics (the study of the nature of sound).

Monge and the Revolution

When the French Revolution broke out in 1789, Monge was one of the few well-known French scientists who was in support of it. Although he was part of the scientific establishment, he also could not forget his modest origins and his commoner status. As a supporter, Monge was soon appointed minister of the navy, a position he would hold for only eight months. His politics would prove to be too moderate for the radicals who were taking over the country,

The Metric System and the French Revolution

The true believers of the French Revolution did not see it as just another political event in history in which different philosophies of government compete and try to assume power or control of the state. Instead, they saw it as a total revolution that would change everything about human society from its politics to its science to the way people lived and even did business. A good example of this is the Revolution's decree that the names of the months be changed and that even the years should start over, beginning with year 1 (naturally the first year of the Revolution). Much of the supporters' reasoning was based on their belief in "rationality" (human reasonableness), science, and the beauty and perfection of everything natural.

As part of Revolution supporters' goal of changing the way children were taught, they decided to come up with a more rational way of counting. They formed a committee, of which Gaspard Monge was a member (along with French mathematicians **Joseph-Louis Lagrange** [1736–1813; see entry in volume 2] and Pierre-Simon de Laplace [1749–1827]), to decide upon a new system of weights and measures. What the committee finally arrived at—the metric system—was probably the most logical system of measurement ever devised and could be described as one of the greatest achievements of the French Revolution.

In 1791, Monge's committee suggested that the Earth itself would be the measuring stick and that the standard unit of length—the meter— should be one ten-millionth of the distance from the equator to the North Pole as it runs through Paris, since this would be the most natural standard or baseline.

The committee also said that measures for **area** (see entry in volume 1) and **volume** (see entry in volume 2) were to be defined in terms of the measures for length. So, the basic unit of mass, the gram, was defined as the mass of a cubic centimeter of water at a given temperature. Today, the metric system is the universal language of science.

Gaspard Monge

Napoléon Bonaparte.
Public domain.

and Monge eventually resigned. However, he continued to support the Revolution, even after it abolished the Académie des Sciences in 1793. In 1791, he helped the Temporary Commission on Weights and Measures to develop the metric system. More importantly, Monge was appointed to a commission that eventually founded the École Centrale des Travaux Publics, which became the École Polytechnique in 1795. Now recognized as the first modern engineering college, Monge was a major influence in setting up and guiding this school.

Link with Napoléon

In 1796, Monge became drawn into the circle of French military commander-in-chief Napoléon Bonaparte (1769–1821), and eventually became one of his friends and advisers. Monge's friendship with a man who would become the dictator of France may seem odd in that Monge was always a champion of the average person and individual freedoms, but many believe that Monge was simply dazzled by the spectacular Napoléon, who gave Monge many honors and titles. In 1796, Monge agreed to go to Rome to "select" the best art treasures and bring them to Napoléon, and in 1798 he accompanied the dictator to Egypt where he did the same thing. By 1809, Monge found his arthritis worse and his health failing, and he resigned his post at the École Polytechnique. As Napoléon's fortunes fell and rose, Monge would sometimes have to flee with his leader, only to be back in power soon after. In several instances, Monge actually had to escape to save his life. With Napoléon's final defeat in 1815, Monge once more returned to Paris, but by 1816, he was being treated badly by those now running France. Although he was not imprisoned, he was expelled from (forced to leave) his scientific associations and harassed for the rest of his life. When Monge died in 1818, the government banned any tributes to him, although some students and friends defied the ban and placed a wreath on his grave the day after his funeral.

A man of humble origins, Monge's mathematical ability allowed him not only to become the founder and creator of descriptive geometry, but to rise to the heights of French political society and eventually come crashing down. Though many believe he made bad political choices, Monge laid the groundwork for the beginning of modern engineering. As historian Eric T. Bell (1883–1960) described him in his *Men of Mathematics*, Monge was "a born geometer and engineer with an unsurpassed gift for visualizing complicated space relations."

For More Information

Bell, Eric T. *Men of Mathematics.* New York: Simon and Schuster, 1937. Reprint, 1986.

Mathematicians and Computer Wizards. Detroit: Macmillan Reference USA, 2001.

O'Connor, J. J., and E. F. Robertson. "Gaspard Monge." *The MacTutor History of Mathematics Archive.* School of Mathe-

Gaspard Monge

Gaspard Monge

matics and Statistics, University of St. Andrews, Scotland. http://www.groups.dcs.st-andrews.ac.uk/~history/ Mathematicians/Monge.html (accessed June 27, 2002).

Taton, René. "Gaspard Monge." In *Biographical Dictionary of Mathematicians*. New York: Charles Scribner's Sons, 1991, pp. 1740–49.

Wilkins, D. R. "Gaspard Monge (1746–1818)." *The History of Mathematics*. From *A Short Account of the History of Mathematics*. W. W. Rouse Ball. http://www.maths.tcd.ie/pub/HistMath/ People/Monge/RouseBall/RB_Monge.html (accessed April 2002).

Born c. 1320
Allemagne, France

Died July 11, 1382
Lisieux, France

French algebraist, geometer, and theologian

Nicole d'Oresme

Although Nicole d'Oresme is considered one of the founders of modern science, he is perhaps best known as a mathematician. His original work helped lay the foundation that later led to the discovery of analytic **geometry** (see entry in volume 1), and overall, it prepared much of the basis for the future development of modern mathematics.

Little known of early life and family

Nicole d'Oresme (pronounced daw-REHM) was born in the Normandy region of France, possibly in the village of Allemagne near Caen (the northwest part of France that faces England across the English Channel). Little is known of his beginnings, but in a document dated 1348, two brothers, Henry and Nicole Oresme, are listed as students of theology (the study of religion) in Paris. Also in 1348, Oresme's name appears on a list of scholarship holders in theology at the College of Navarre in Paris. It was at this college that he met and became friends with the son of King John II (1319–1364), Prince Charles (1337–1380), who later became King Charles V. Before this, Oresme attended the University of Paris in the 1340s, where he studied with noted theologian, philosopher, and physicist

Jean Buridan (pronounced byoo-ree-DAHN) (1300–c. 1358). At the University of Paris, Oresme's studies included the liberal arts and theology, which qualified him to teach theology.

Nicole d'Oresme

In the Middle Ages (the time between the fall of Rome and the Renaissance, which ended in the sixteenth century), priests or clergy usually ran the universities and many of their students were expected to enter the priesthood. However, the more talented or gifted students went beyond this and received a liberal arts education. In the fourteenth century, this meant that their studies consisted of the *trivium* (the courses of grammar, rhetoric, and logic), and the *quadrivium,* (the courses of arithmetic, music, geometry, and astronomy). Following this, the best students continued to study philosophy, which had four branches: theoretical, practical, logical, and mechanical. Since Oresme remained in school, he was evidently a talented individual and he became very capable in all of these fields.

Increasingly responsible positions

Oresme received his doctorate in 1356 from the College of Navarre, and became headmaster (principal) there the same year. He was soon appointed bursar (pronounced BER-ser; a position similar to a treasurer in which a person controls funds). In 1362, Oresme left Navarre to become a canon (a type of clergyman) at the Cathedral in Rouen. That same year, his friend Prince Charles became King Charles V, and the new king appointed Oresme to be his chaplain and counselor (financial adviser). Two years later, Oresme was named dean at Rouen. In 1372, he was made the bishop of Lisieux (in Normandy), France, and it would be in this city that he would spend his final years.

Oresme lived during what many consider the most advanced part of the Middle Ages. He was strongly influenced by the rational and empirical (based on experience or observation) spirit of his time. Many of the leading thinkers of his age argued that the only true foundation of knowledge was actual experience, as opposed to accepting something because it always has been. Such an attitude toward knowledge is important for a person to discover something new.

Mathematical contributions

One of Oresme's most important contributions to mathematics was a graphing system he devised. He applied the system to the

Sister Mary Celine Fasenmyer

Sister Mary Celine Fasenmyer.
Reproduced by permission of A. K. Peters Ltd.

If it seems surprising that Nicole d'Oresme was both a highly original mathematician and a medieval bishop in the Roman Catholic Church, the case of Sister Mary Celine Fasenmyer (1906–1996) may be even more unusual. While it was highly unlikely that a woman born at the turn of the century to a working-class family in a small oil town in central Pennsylvania would be able to attend the University of Pittsburgh and earn a Ph.D. in mathematics, it is even more startling to learn that this woman had been a Roman Catholic nun or sister since the age of seventeen.

Fasenmyer, or Sister Celine as she was known, had the good fortune to join the Sisters of Mercy, which was a teaching order. This not only meant that most of the young women who joined this order became teachers, but that the order itself knew the value of and had great respect for higher education. Since Sister Celine showed more than average ability in mathematics, she was allowed to take graduate studies after she received her bachelor's degree. The hard-working nun finally received her doctorate from the University of Michigan in 1946 at the age of forty. While this alone was a major achievement, it was what she did while getting her advanced degree that allowed her to really make her mark as a mathematician. Sister Celine's specialty was algorithms, a mathematical term for any step-by-step procedure or systematic method that is used to solve a mathematical problem. Besides her doctoral dissertation, she published only two papers in her entire career, one in 1947 and another in 1949. Yet it is on the basis of this remarkably small output that she is now recognized as being the "mother" of the computerized methods used today to prove hypergeometric identities. (Hypergeometric refers to being able to prove recurrence relations, or the pattern or probability of things.) Sister Celine laid the intellectual groundwork that enabled others to discover how to use computers to simplify highly complex equations. Using her methods, other mathematicians were able to detect patterns that were nearly impossible for them to discover using traditional methods.

Sister Celine's work was not used until very late in her life, and she spent her entire career teaching at Mercyhurst, a Catholic college in Erie, Pennsylvania. One of the men who became famous using her approach, University of Pennsylvania mathematics professor Herbert S. Wilf (1931–), finally met Sister Celine in 1994 when he went to her Pennsylvania retirement home and took her to a mathematics conference in Florida in which her work and her accomplishments were formally recognized. The 87-year-old nun spoke briefly, saying, "I want you all to know—I really did that work."

Nicole d'Oresme

problem of an object that is moving at a uniform rate for a fixed period of time, and how much distance it will cover. His idea was to plot this problem on a two-dimensional graph. In order to represent an object's velocity or speed as it changes over time, he represented time along a horizontal line, which he called the longitude, and represented the velocities at various times by vertical lines, which he called latitudes. Although this idea was not entirely original to Oresme—it had not been thought of or used since the time of the ancient Greeks—it was certainly a new idea to mathematics of his time. In using a graphing system to work on the study of motion (velocity, time, and distance), Oresme made the important breakthrough in showing that there were measurable quantities other than numbers that could be represented by points, lines, and surfaces. He, thus, was able to demonstrate the problem with the geometrical figure that he made with the graph, and numerical values or numbers had nothing to do with it.

Oresme's new system allowed him and others to study motion theoretically (without having to do any actual experiments). It was because of this that his work is considered to be a precursor (something that comes before) to modern analytic geometry. Also called **coordinate graphing** (see entry in volume 1), analytic geometry can be described as the application of **algebra** (see entry in volume 1) to the study of geometry. Some three hundred years later, French algebraist, geometer, and philosopher **René Descartes** (1596–1650; see entry in volume 1), would reinvent analytic geometry and refine it, but it is not known whether Descartes knew of Oresme's work.

Oresme also undertook a detailed study of **ratios** (see entry in volume 2) and made important contributions in understanding this mathematical concept. A ratio shows the comparison of two related numbers, quantities, or terms. Speaking strictly mathematically, a ratio is the relationship obtained by dividing things. Put yet another way, it is the comparison of two quantities by **division** (see entry in volume 1). Oresme found that one also can compound or increase ratios by multiplying the antecedent (the first term of a ratio) and then multiplying the consequent (the second term of a ratio). (In the ratio 4 : 3, the antecedent is 4 and the consequent is 3.) In other words, 4 : 3 (4 to 3) compounded with 5 : 1 (5 to 1) is the same as 20 : 3 (20 to 3). Oresme eventually

A page from a manuscript by Aristotle that Nicolas d'Oresme translated. In the upper left, an illustration shows Oresme presenting his translation to King Charles V of France. *Public domain.*

was able to develop an arithmetic for these and other ratios, and in many ways, his work shows for the first time some operational rules for dealing with ratios.

Questions geocentric system

One of the issues related to the study of motion that interested Oresme was the notion of the Earth's movement. Oresme argued persuasively that the Earth was in motion some two hundred years before Polish astronomer Nicolas Copernicus (1473–1543) stated his revolutionary theory that the Earth orbited the Sun. Most people of that time believed in the geocentric system, which stated that the Sun orbited the Earth. Oresme pointed out that what people think of as the heavens moving around a stationary Earth may be only people's senses deceiving them. He even questioned the Bible, which refers to the Sun's rotation around the Earth.

Oresme wrote a good deal; one of his last writings was his translation of some of the works of Greek philosopher Aristotle

Nicole d'Oresme

(384–322 B.C.E.), which was commissioned by Charles V. But, unfortunately, Oresme lived in the century before the invention of mechanical printing, and most of his writings remained in hand-written form, and were, therefore, not easily or quickly reproduced and made available. But Oresme's reputation as one of the founder's of modern science remains intact.

For More Information

Clagett, Marshall. "Nicole Oresme." In *Biographical Dictionary of Mathematicians*. New York: Charles Scribner's Sons, 1980, pp. 223–30.

Duhem, Pierre. "Nicole Oresme." In *The Catholic Encyclopedia*. Online edition by Kevin Knight. http://www.newadvent.org/cathen/11296a.htm (accessed June 28, 2002).

Franceschetti, Donald R., ed. *Biographical Encyclopedia of Mathematicians*. New York: Marshall Cavendish, 1999.

Kline, Morris. *Mathematical Thought from Ancient to Modern Times*. New York: Oxford University Press, 1972, pp. 210–11.

O'Connor, J. J., and E. F. Robertson. "Mary Celine Fasenmyer." *The MacTutor History of Mathematics Archive*. School of Mathematics and Statistics, University of St. Andrews, Scotland. http://www.groups.dcs.st-andrews.ac.uk/~history/Mathematicians/Fasenmyer.html (accessed June 28, 2002).

O'Connor, J. J., and E. F. Robertson. "Nicole d'Oresme." *The MacTutor History of Mathematics Archive*. School of Mathematics and Statistics, University of St. Andrews, Scotland. http://www.groups.dcs.st-andrews.ac.uk/~history/Mathematicians/Oresme.html (accessed June 28, 2002).

Wilf, Herbert. "Sister Mary Celine Fasenmyer." *Biographies of Women Mathematicians*. Agnes Scott College. http://www.agnesscott.edu/lriddle/women/celine.htm (accessed June 28, 2002).

**Born June 19, 1623
Clermont (now Clermont-Ferrand), France**

**Died August 19, 1662
Paris, France**

French geometer and physicist

Blaise Pascal

I n his short life, Blaise Pascal made several contributions to
mathematics. He opened up new forms of calculus and projective **geometry** (see entry in volume 1) and helped develop the
foundations of **probability** (see entry in volume 2) theory. He
also designed and built the first calculating machine, run by cogs
and wheels.

Blaise Pascal.
*Reproduced by permission of the
Corbis Corporation.*

Child prodigy

Blaise Pascal (pronounced pahs-KAHL) was born in the city
now known as Clermont-Ferrand, France. He was the son of
Etienne Pascal, a skilled mathematician and a high-ranking civil
servant, and Antoinette Begon. He had two sisters, Gilberte and
Jacqueline. Pascal's mother died when he was only three years
old, and five years later his father moved the family to Paris.
Young Pascal was part of a very close family, and he was especially close to his two sisters. Pascal was a prodigy (a child of exceptional intelligence and ability), and his father decided he would
educate him at home himself. His plan was to first have his son
study ancient languages and only later would he introduce him
to mathematics. Pascal was very curious about geometry, but his

Blaise Pascal

father would not teach him this subject until he was older. However, young Pascal's curiosity continued, and he began to try to figure out geometry on his own, without any instructions or books. Incredibly, by the time he was eleven, he had discovered for himself the writings of Greek geometer **Euclid of Alexandria** (c. 325–c. 270 B.C.E; see entry in volume 1). When his father learned of this, he gave in and agreed to teach him all the mathematics he knew.

Rapid mathematical progress

In Paris, Pascal's father had begun attending meetings of a group of intellectuals, usually scientists and mathematicians. This group called itself the Académie Parisienne. One of the members of this group was French geometer, algebraist, and philosopher **René Descartes** (1596–1650; see entry in volume 1). Young Pascal benefited greatly from his exposure to such intellects, and by the time he was sixteen years old, he had already completed a significant amount of difficult mathematical work. Although his father moved the family again in 1639, this time to Rouen when he was appointed to the tax office there, his son continued to make trips to Paris to meet with this stimulating group.

In 1646, at the age of seventeen, Pascal published a pamphlet, *Essai sur les coniques* (Treatise on Conic Sections), which he presented to the group. This mathematical essay was so advanced and impressive that Descartes at first refused to accept it as being the work of such a young person. This essay would prove to be an important step in projective geometry. Projective geometry is different from the traditional form of geometry taught by Euclid. It was called "projective" because it involves the study of the properties related to transferring a three-dimensional object to a two-dimensional surface (such as a painter might do). Pascal took up this subject to clarify the work of French geometer and engineer **Girard Desargues** (1591–1661), who had published a book on the subject in 1639. Since Desargues's book was written in a very difficult style, Pascal first tried to simplify the work, but then began to experiment and ended up going beyond what Desargues had accomplished. Pascal eventually developed his own theorem from which he worked out some four hundred corollaries (results of something that naturally follows from something else).

Develops calculating machine

Soon after Pascal's pamphlet was published, he began work on what was in effect the first digital calculator. He had begun work on this project in part to help his father who had to perform lengthy calculations at work in his tax office in Rouen. By 1644, he was able to demonstrate his calculating machine, which could automatically add and subtract using cogged (jagged, teeth-like) wheels to do the calculations. This was a significant invention for anyone of any age, but for someone of his young age it was truly remarkable. Pascal patented his invention and received exclusive ownership by a royal decree. Although he wanted to manufacture these machines as a full-scale business enterprise, this proved to be too expensive. Still, the basic principle behind Pascal's calculator continued to be used up until the twentieth century, before computers went electric.

Health problems and new religion

Not long after he published his pamphlet, Pascal's health began to decline. He suffered from severe headaches, an inability to fall asleep, and indigestion. From about 1641 to the end of his life, he

The mechanical calculating machine, invented by Blaise Pascal. *Reproduced by permission of Corbis-Bettmann.*

Blaise Pascal

would never really enjoy good health. In 1646, however, his life changed considerably when he became part of a separate Catholic movement called Jansenism. This group, named after Dutch theologian Cornelius Jansen (1585–1638), believed in predestination (the idea that everything has been determined beforehand) and that the only way to be "saved" was by receiving God's grace or blessing. Pascal felt so strongly about his new faith that he persuaded his entire family to join him, and Jansenism would play a dominant role in the rest of his life.

Physics experiment

About this same time, Pascal began conducting numerous experiments concerning vacuums as well as atmospheric (relating to the entire mass of air surrounding Earth) and barometric (the pressure of the atmosphere) pressure. Italian physicist Evangelista Torricelli (1608–1647) had recently discovered the principle of the barometer, which measures air pressure, and Pascal decided to test his theories. Pascal used mercury barometers to measure air pressure in Paris and at the top of a nearby mountain called Puy de Dôme. Since Pascal was too frail to do this himself, he had his brother-in-law carry the mercury barometer to the summit and get readings. What he discovered was that the column of mercury in the tube rose three inches with the increase in altitude. This meant that if air has weight and presses down on the mercury, then that liquid ought to rise in the column since there would be less air pushing down on it at a mountaintop than there would be much farther below (in Paris). These experiments led to his discovery of what came to be called Pascal's principle, that pressure applied to a confined liquid (like mercury in a tube) is transmitted through the liquid in all directions regardless of the area over which the pressure is applied. His work also led to his practical development of the syringe (pronounced suh-RIHNJ; a device that injects or withdraws fluid) and the hydraulic (pronounced hy-DRAW-lik; the energy effects of moving water) press. Pascal published some of his results in 1647 and 1648.

Foundations of probability

In 1654, Pascal began a correspondence with French number theorist **Pierre de Fermat** (1601–1665; see entry in volume 1), in which he wrote and asked him for advice on a problem involving the consecutive throws of a die. Pascal asked Fermat how to divide the stakes in a dice game between two players when their game was prematurely interrupted. This question caught Fermat's interest, and over the next few months the letters they exchanged would help lay the mathematical foundations of probability theory (the study of the likely outcomes that an event will occur). Their work would eventually show the importance of what appeared to be random, unpredictable acts.

Religious experience and last work

In 1647, the Pascal family returned to Paris, and it was there that his father died in 1650. By 1654, Pascal was alone as his sister Gilberte had married, and Jacqueline had entered the Jansenist convent in Port-Royale. On the night of November 23, 1654, Pascal experienced what he considered to be a profound religious experience. That night, after he was nearly killed in a riding accident, Pascal said he had some sort of mystical experience of God and Jesus Christ, which made him decide to give up all science and devote himself to his religious studies. He then moved to the Jansenist retreat in Port-Royale and devoted himself to God.

Pascal did his last mathematical work in 1658, four years after his religious experience. He was suffering from a painful tooth and decided to try to distract himself by working out an old problem concerning the geometry of a cycloid (pronounced SYE-kloyd). A cycloid is the name of the curve that is traced by the motion of a fixed point on the circumference of a **circle** (see entry in volume 1) rolling along a straight line. Over the course of eight days of intense concentration, Pascal solved many of the remaining unknowns about the geometry of the cycloid. This work would eventually play a major role in the development of calculus.

Literary work

Between his religious experience in 1654 and his death in 1662, Pascal wrote and anonymously published eighteen pamphlets that altogether were known as *Les Provinciales* (The Provincial Letters). These essays were intended to sway public opinion in favor of the Jansenists and against their religious opponents called the Jesuits (pronounced JEH-zoo-itz). These were an immediate success as he brilliantly criticized and made fun of his opponents' religious

Blaise Pascal

positions. He also wrote but did not complete an overall argument in favor of Christianity. When these were published after his death in 1670 as *Pensées* (Thoughts), they too became very popular. In these essays, Pascal sees humans as being created by God for some sort of greatness but as being at the same time lost in a vast universe in which they can do nothing without God. Pascal died at the age of thirty-nine in his sister Gilberte's house, probably of stomach cancer.

For More Information

Franceschetti, Donald R. *Biographical Encyclopedia of Mathematicians.* New York: Marshall Cavendish, 1999.

Mathematicians and Computer Wizards. Detroit: Macmillan Reference USA, 2001.

McPherson, Joyce. *A Piece of the Mountain: The Story of Blaise Pascal.* Lebanon, TN: Greenleaf Press, 1997.

O'Connor, J. J., and E. F. Robertson. "Blaise Pascal." *The MacTutor History of Mathematics Archive.* School of Mathematics and Statistics, University of St. Andrews, Scotland. http://www.groups.dcs.st-andrews.ac.uk/~history/Mathematicians/Pascal.html (accessed June 28, 2002).

Taton, René. "Blaise Pascal." In *Biographical Dictionary of Mathematicians.* New York: Charles Scribner's Sons, 1991, pp. 1914–25.

Wilkins, D. R. "Blaise Pascal (1623–1662)." *A Short Account of the History of Mathematics.* School of Mathematics, Trinity College, Dublin. http://www.maths.tcd.ie/pub/HistMath/People/Barrow/RouseBall/RB_Pascal.html (accessed June 18, 2002).

Born c. 100
Egypt

Died c. 170
Alexandria, Egypt

Greek geometer, astronomer, and geographer

Claudius Ptolemy

T‌he astronomical and geographical work of Claudius Ptolemy dominated Western thought for nearly fourteen hundred years. He developed new geometrical proofs and theorems to back his astronomical theories and made lasting contributions to the field of trigonometry. He also introduced the system of latitudes and longitudes in the use of modern maps and charts.

Claudius Ptolemy.
Courtesy of the Library of Congress.

Nothing known of personal life

Almost the entire life of Claudius Ptolemy (pronounced TAHL-uh-mee) remains a mystery outside of his publications, but researchers have pieced together some clues of his childhood and education. His name sounds a bit strange in that it combines names that are well-known both in ancient Rome and Egypt. Ptolemy was a common name in Egypt, leading most to believe that he was a native of that country. His first name, Claudius, suggests that he held Roman citizenship, probably due to a legacy handed down through his Greek relatives, the result of a citizenship granted to an ancestor by a Roman emperor. From this information, historians conclude that he was descended from a Greek family living in Egypt, and that he was a Roman citizen. Thus, his

Claudius Ptolemy

full name in Latin would have been Claudius Ptolemaeus. The best evidence scholars have about his life is based on his own recorded astronomical observations, all of which were made in Alexandria, Egypt. This is the only place mentioned in any of his observations. The dates of these observations are used to approximate the dates of his life. Specifically, the first of these was made exactly on March 26, in the year 127, and the last was made on February 2, in the year 141.

Ptolemy almost certainly lived in Alexandria, Egypt, where he made his astronomical observations. That city's famous library provided him with access to the largest collection of accumulated knowledge in the world. From his writings, scholars learn that he probably studied the work of Greek philosopher Aristotle (384–322 B.C.E). Ptolemy's writings also illustrate that he was, however, influenced by other Greek philosophies such as stoicism (the belief that the universe is basically rational or understandable).

Advances trigonometry through astronomy

Ptolemy's long-lasting fame is based on his large, multivolume work titled *Almagest* (The Greatest). As the earliest of Ptolemy's works, it is basically a manual covering the whole of mathematical astronomy. Written in Greek, its original title was *Mathematike syntaxis* (Mathematical Collection). This work eventually became so admired and gathered so much prestige that when it was translated into Arabic from the Greek, it was given the title *al-Majisti* (Great Work). Later, when it was translated from the Arabic into Latin, the name became *Almagesti* or *Almagestum*, and later simply *Almagest*.

This mathematical collection was a summary of Greek astronomy to that point in time, and it became the main source for information on a great Greek astronomer, Hipparchus (fl. 146–127 B.C.E.; pronounced hih-PAHR-kus). Because of this, many are not clear as to which ideas belong to Ptolemy and which came from Hipparchus. However, since it is known that Ptolemy added some 850 stars to about 200 already catalogued by Hipparchus, it is apparent that Ptolemy certainly did a considerable amount of work on his own. Ptolemy's *Almagest* provides in great detail the mathematical theory of the motion of the Sun, Moon, and planets. Further, Ptolemy described a mathematical arrangement of the stars and even gave each one a latitude (the horizontal lines on a map), longitude (the vertical lines on a map), and magnitude (brightness rating). This in itself was a major contribution as it introduced the use of latitudes and longitudes, both of which are a means of measuring distance and locating a particular spot on a map in relation to a reference point.

Ptolemy's work is important not only as a manual of astronomy but because it is a mathematical treatment of astronomy. Since the *Almagest* introduces the use of trigonometry (the study of the sides and angles of triangles and their measurements and relations), Ptolemy is considered as having greatly advanced the study of trigonometry (although Hipparchus is considered to be the founder of trigonometry). Ptolemy used trigonometry to make various calculations for solar, lunar, and planetary positions as well as for eclipses of the Sun and the Moon. Perhaps most important, Ptolemy's work provided a geocentric model for astronomers to follow. Called the geocentric model because it placed the Earth at the center of the universe, this theory stated that all of the heavenly bodies orbited or rotated around a stationary (unmoving) Earth.

Ptolemy also created a sophisticated mathematical model to fit his "data." He believed that the Earth was round but did not move and was located at the center of the universe. Since he knew from experience that all heavy objects fall to the ground—and therefore, he said, they want to fall to the center of the Earth—he argued that it made sense that all the planets, stars, and even the Sun and Moon either move toward the Earth or around it. His model, called the Ptolemaic system, also stated that the planets move at different speeds, each in its own particular and perfect circular orbit. To make his elaborate system work and be able to predict future planetary positions, he added an ingenious system that used combinations of circles, known as epicycles, or small circular orbits centered on a larger circle's circumference. The geocentric model, of course, would eventually be proven wrong.

Establishes scientific foundation for geography

Ptolemy's second great scientific work was his *Geographike hyphegesis* (Guide to Geography). Written in eight sections, this work is an early attempt to map the world, and although much of it is inaccurate, it nonetheless placed the geographical knowledge accumulated by the Greeks and Romans on a solid scientific foundation. In this

A print entitled "Harmonia Macrocosmica" shows the Ptolemaic system.
Print by Andea Cellario. Photograph by Enzo and Paolo Ragazzini. Reproduced by permission of the Corbis Corporation.

work, Ptolemy attempted to map the known world (Europe, northern Africa, and most of Asia) by giving the **coordinates** (see entry on coordinate graphing in volume 1) of its major places in terms of latitude and longitude. The latitude and longitude directions for mapmaking that Ptolemy included in this work were followed by geographers until the Renaissance (the fifth century until the end of the seventeenth century). Unfortunately, Ptolemy underestimated the circumference of the Earth, leading later navigators to sometimes be seriously led astray. It is believed that Italian explorer Christopher Columbus (1451–1506) relied on Ptolemy's world map when he planned his expedition west. Since this map made the land masses of Europe and China too large and the ocean too small, Columbus sailed westward expecting to reach Asia much sooner than he did. As is now known, however, Columbus was unable to reach Asia since there was another huge, unknown continent (North America) in between Asia and Europe. Despite this and many other errors, Ptolemy's *Geographike hyphegesis* set the scientific standard for maps and was influential to the time of Columbus and beyond. Another of its innovations was Ptolemy's establishment of the cartographic (relating to mapmaking) principle of placing north at the top of a map. Besides astronomy and geography, Ptolemy also wrote on the theory of music and optics (the science of light), and believed in astrology (the belief that the stars and planets influence human affairs).

The astronomical teachings of Ptolemy were so dominant, that his ideas were considered the ultimate authority until the end of the sixteenth century. It did not hurt, however, that his geocentric or Earth-centered model of the universe was accepted by Christianity and was taught as an unchallenged truth. Since the Bible stressed the overwhelming importance of the Earth as a creation of God's and the place where he located the humans he also created, Ptolemy's model, which placed the Earth at the center of the universe, seemed to emphasize and make scientifically legitimate the unique significance of the Earth to the universe. Although there have been some who have criticized his ability and skills, most believe that he was indeed a mathematician of the first rank, especially when one considers how sophisticated (although incorrect) were the mathematical models he created to fit his observational data.

For More Information

Ellis, Walter M. *Ptolemy of Egypt*. London: Routledge, 1994.

Fransceschetti, Donald R., ed. *Biographical Encyclopedia of Mathematicians*. New York: Marshall Cavendish, 1999.

Mathematicians and Computer Wizards. Detroit: Macmillan Reference USA, 2001.

Toomer, G. J. "Ptolemy." In *Biographical Dictionary of Mathematicians*. New York: Charles Scribner's Sons, 1991, pp. 2053–73.

O'Connor, J. J., and E. F. Robertson. "Claudius Ptolemy." *The MacTutor History of Mathematics Archive*. School of Mathematics and Statistics, University of St. Andrews, Scotland. http://www.groups.dcs.st-andrews.ac.uk/~history/Mathematicians/Ptolemy.html (accessed June 30, 2002).

Claudius Ptolemy

Claudius Ptolemy

Ptolemy, Claudius. *Ptolemy's Almagest*. Translated and annotated by G. J. Toomer. New York: Springer-Verlag, 1984.

**Born December 8, 1919
St. Louis, Missouri**

**Died July 30, 1985
Berkeley, California**

American logician and number theorist

Julia Bowman Robinson

O ne of the first great American women mathematicians, Julia
Bowman Robinson helped find the solution to one of the
famous unsolved problems of mathematics. In recognition of this and
other accomplishments, she became the first woman elected to the
mathematical section of the National Academy of Sciences, as well as
the first woman president of the American Mathematical Society.

Julia Bowman Robinson.
*Reproduced by permission of
Mathematisches
Forschungsinsitut Oberwolfach.*

Health problems did not affect her love of math

Julia Bowman Robinson never really knew her mother, Helen Hall
Bowman, who died when she was only two years old. Soon after,
Julia and her older sister, Constance, left St. Louis, Missouri, and
went to live with their grandmother near Phoenix, Arizona. The following year, their father, Ralph Bowman, remarried and joined them
in Arizona. Julia's father had already sold his machine tool and equipment business and planned to live on his savings and his invest-
ments. The entire family then moved to San Diego, California, in
1925 and Ralph Bowman and his new wife had a baby of their own.

As a youngster, Robinson was considered "slow" since she was
both a stubborn child and a very late talker. When she did finally

speak, she was difficult to understand, so her sister Constance often did much of the talking for her. Both were in the same classroom, since the school they attended was so small that it had several different grades together in one room. When Robinson was nine, she contracted scarlet fever, which was a very serious disease before the emergence of antibiotics. This infectious disease is caused by bacteria and frequently has serious complications. Unfortunately for Robinson, she then developed rheumatic fever, another very serious condition, which often damages the heart and, after several relapses, she was forced to stay in bed for an entire year. As a result, Robinson lost more than two years of school and more seriously, permanently damaged her heart.

Throughout her bed rest period, Robinson worked with a tutor and made up all the course work and subjects covered in grades five through eight. Her real interest in mathematics began around that time; she became fascinated when her tutor told her that if she tried to work out the **square root** (see entry in volume 2) of two, the result was a **decimal** (see entry in volume 2) number that never ended and never repeated. She had to see this for herself, and she once spent an entire afternoon calculating one digit after another. In 1932, when she was finally healthy enough to attend school, she entered the ninth grade in San Diego. Since she had been away so long, she was somewhat shy so she concentrated very hard on doing well in school, especially in mathematics. She enjoyed keeping box scores of baseball games, and especially liked horseback riding and target practice with her father. In high school, she excelled at the basic courses of **algebra** (see entry in volume 1), trigonometry, and **geometry** (see entry in volume 1), after which she became the only girl taking advanced mathematics and physics courses. She later recalled that being the only girl in a roomful of boys never bothered her, and that the boys only paid attention to her when they needed help. In fact, she did so well in school that she not only caught up, but also graduated when she was only sixteen. Upon graduation, she received awards in every scientific subject she took, and also received a special medal from the Bausch & Lomb Company for all-round excellence in science and math.

Decides to become a mathematician

When Robinson entered San Diego State College in 1936, she knew she wanted to major in mathematics. She had no real notion that a person could be a full-time mathematician, and she assumed that her career choices in math were limited to teaching the course in high school. However, a book she read, called *Men of Mathematics,* by Eric T. Bell (1883–1960), opened up a whole new world for her. It showed her that there were still a great number of very interesting mathematical problems left to solve, and it exposed her to people whose lives revolved around mathematics. After reading the book, Robinson decided she would give up simply trying to be a teacher and would do whatever was necessary or go wherever she had to in order to learn to be a mathematician.

However, just as Robinson was deciding that she should transfer to the University of California at Berkeley, her father committed suicide. His life savings had been wiped out in the economic depression that followed the stock market crash of 1929. Robinson was determined to overcome this setback, and with the help of her older sister and an aunt, she transferred to Berkeley for her senior year. With that, her life changed as she suddenly found herself surrounded by students who were as interested and excited by mathematics as she.

Mathematics becomes her life

Robinson graduated from Berkeley in 1940 and became a teaching assistant there while she studied for her master's degree. During this time, she and a Berkeley mathematics professor, Raphael M. Robinson, began spending more time together, and they were married on December 22, 1941. Her marriage would limit her career opportunities, however, since she could not continue in her teaching job at the university because of its rule prohibiting a husband and wife from working in the same academic department. She took a job instead in the Berkeley Statistical Laboratory working on military projects during World War II (1939–45). She and her husband had always planned to have a family, and when she became pregnant they were both very happy. But Robinson suffered a miscarriage and lost the baby. Her doctors then discovered that the rheumatic fever had caused a build-up of scar tissue in a heart valve, and that she should not become pregnant again. One doctor also theorized privately that Robinson would be lucky to live to age forty.

Losing a baby and finding out that she could never have children caused her to become deeply depressed, but with her husband's help, Robinson found in her mathematics a reason to go on living. Robin-

Julia Bowman
Robinson

son eventually began work on her Ph.D. at Berkeley under the direction of Alfred Tarski (1902–1983), a Polish mathematician who she believed to be one of the two most important logicians (those who study reasoning) of the twentieth century, the other being Austrian American **Kurt Gödel** (1906–1978; see entry in volume 1). Under Tarski's direction, Robinson obtained her doctorate in 1948.

Solves Hilbert's "10th Problem"

After receiving her Ph.D., Robinson took a research position with the RAND Corporation and also became interested in a mathematical problem that German number theorist David Hilbert (1862–1943) had posed near the beginning of the century. In his lifetime, Hilbert was one of the best-known mathematicians in the world, and in 1900 he compiled a list of twenty-three challenges to other mathematicians of what were then the most important unsolved problems. The tenth problem on his list had to do with whether anyone could ever determine if a particular type of equation could be solved. The background of this problem actually went all the way back to the seventeenth century and French number theorist **Pierre de Fermat** (1601–1665; see entry in volume 1).

Throughout her career, Robinson worked on the problem little by little and published several papers. She said she "sneaked up" on the problem by working on other problems, which, if solved, would become steps toward a complete solution. Every year, when she blew out her birthday candles, she would wish that the tenth problem would be solved in her lifetime. In 1961, she underwent a risky surgical procedure on her heart that proved successful. With renewed energy (which even led her to take up serious bicycle riding), Robinson kept working on the Hilbert problem. Finally, in February 1970, she received a phone call from a colleague who said that a young Russian mathematician had used her hypothesis (an explanation that accounts for a set of facts) and arrived at an answer.

Humbly accepts honors

Indeed, a twenty-two year old mathematician named Yuri Matiyasevich (mah-tih-YAH-seh-vitch; 1947–) stated that it was Robinson's work that led to his ultimate solution, and he credited her work as the foundation of his. Robinson congratulated him, say-

Julia Bowman Robinson

Julia Bowman Robinson spent much of her career trying to solve the tenth of twenty-three mathematical challenges that German number theorist David Hilbert (above) compiled in 1900.
Reproduced by permission of the Corbis Corporation.

ing, "If you really are twenty-two, I am especially pleased to think that when I first made the conjecture you were a baby and I just had to wait for you to grow up." She and her husband later visited him in Leningrad. Many mathematicians stated that the solution to the problem followed so naturally from Robinson's work, they wondered why she had not completed the solution herself.

In recognition of her life's work and accomplishments, Robinson was made a full professor at Berkeley in 1975. That same year, she became the first woman elected to the mathematical section of the National Academy of Sciences. In 1983, she also became the first woman elected to the presidency of the American Mathematical Society. By this time, her health had weakened again, and although she knew she was not physically up to the hard work the presidency involved, she said that "as a woman and as a mathematician I had no alternative but to accept. I have always tried to do everything I could to encourage talented women to become research mathematicians." Robinson then put aside her shyness and poured her last energy into this post. It was during the society's annual conference that she learned she was suffering from leukemia. After bravely battling this serious blood disease for almost a year, Robinson died at the age of sixty-five. Her life was longer and more productive than she or her doctors ever could have imagined.

For More Information

Bricker, Julie. "Julia Bowman Robinson." *Biographies of Women Mathematicians.* http://www.agnesscott.edu/lriddle/women/robinson.htm (accessed June 30, 2002).

"Julia Bowman Robinson, 1919–1985." *Notices of the American Mathematical Society* (November 1985): 738–42.

Matiyasevich, Yuri. "Julia Bowman Robinson on the Internet." *Welcome to WWW Pages of Yuri Matiyasevich!* http://logic.pdmi.ras.ru/~yumat/JRobinson/index.html (accessed June 30, 2002).

Matiyasevich, Yuri. "My Collaboration with Julia Robinson." *The Mathematical Intelligencer* (Fall 1992): pp. 38–42.

Morrow, Charlene, and Teri Perl, eds. *Notable Women in Mathematics: A Biographical Dictionary.* Westport, CT: Greenwood Press, 1998.

O'Connor, J. J., and E. F. Robertson. "Julia Bowman Robinson." *The MacTutor History of Mathematics Archive.* School of Mathematics and Statistics, University of St. Andrews, Scotland. http://www.groups.dcs.st-andrews.ac.uk/~history/Mathematicians/Robinson_Julia.html (accessed June 30, 2002).

Reid, Constance. "The Autobiography of Julia Robinson." *The College Mathematics Journal* (January 1986): pp. 2–21.

Reid, Constance. "Being Julia Robinson's Sister." *Notices of the American Mathematical Society.* (December 1986): pp. 1486–92. http://www.ams.org/notices/199612/reid.pdf (accessed June 30, 2002).

Reid, Constance. *Julia, a Life in Mathematics.* Washington, DC: Mathematical Association of America, 1996.

Reid, Constance, with Raphael M. Robinson. *Women of Mathematics.* Edited by Louise S. Grinstein and Paul J. Campbell. Westport, CT: Greenwood Press, 1987, pp. 182–89.

Smorynski, C. "Julia Robinson: In Memoriam." *The Mathematical Intelligencer* (Spring 1986): pp. 77–79.

Julia Bowman Robinson

Born March 18, 1796
Utzenstorf, Switzerland

Died April 1, 1863
Bern, Switzerland

Swiss geometer

Jakob Steiner

Jakob Steiner.
Reproduced by permission of
Mathematisches Forschungsinstitut
Oberwolfach.

Jakob Steiner is considered not only the father of modern projective **geometry** (see entry in volume 1), but the greatest geometer of modern times. Several geometric theorems and principles are named after his discoveries, and his work is regarded as the classical authority for projective geometry.

Learns to read as teenager

Jakob Steiner (pronounced SHTYNE-ur) was born in Utzenstorf, near Bern, Switzerland. He was the youngest of eight children of his mother, Anna Barbara Weber, and his father, Niklaus Steiner, a small farmer and a tradesman. Steiner grew up without any formal schooling since he had to work on the family farm and in his father's business. Although he displayed a natural skill with numbers, he did not learn how to read or write until he was fourteen years old. Steiner had a natural desire for learning, so when he turned eighteen he left home against his parents' wishes in order to get more of an education.

Steiner went west to Yverdon, a city near the French border, to attend a school run by Swiss educational reformer Johann Hein-

Jakob Steiner

Johann Heinrich Pestalozzi, whose educational methods were embraced by Swiss geometer Jakob Steiner.
Reproduced by permission of Archive Photos, Inc.

rich Pestalozzi (1746–1827). Established in 1805, students from all over Europe attended this school, and it was there that Pestalozzi was able to employ his theories that laid the foundation for modern elementary education. The Pestalozzi method took into account the individual needs of each student and stressed the importance of accurate observations. His method also rejected strict memorization of facts in favor of actual experience and critical thinking. Steiner took to this method so well, and was so bright, that after eighteen months with Pestalozzi, he was made a

Jakob Steiner

teacher of mathematics at the school. Pestalozzi found that young Steiner was both a brilliant example and an excellent interpreter of his own revolutionary ideas. This experience of uncovering one's natural ability and of thinking independently would always stay with Steiner, especially when he dealt with his own students.

Establishes his career

In 1818, Steiner left Pestalozzi's school and traveled to Heidelberg, Germany, where he supported himself by giving private instruction in mathematics. Since he was earning a modest living, he was able to attend mathematical lectures at the university. This experience gave him enough skill and knowledge to be able to leave Heidelberg in 1821 for the Prussian capital of Berlin, where he became a student at the University of Berlin. After attending that school from 1822 to 1824, he taught at a technical school in Berlin until 1834, when he was appointed extraordinary professor at the University of Berlin. He would hold this position until his death.

As early as 1821, Steiner wrote the first of his many mathematical papers, and in 1826, he began to conduct his original research. Almost from the beginning of his mathematical education, Steiner showed a passion for geometry and eventually made reforming (changing) geometry his life's goal. His 1826 article was published in a new mathematical journal founded by his friend, an influential engineer, August Leopold Crelle (1780–1855). First published in 1825, the *Journal for Pure and Applied Mathematics,* commonly known as *Crelle's Journal,* would become the leading German mathematical periodical of the nineteenth century and is still published in modern time. Altogether, Steiner would contribute sixty-two articles to that journal.

Projective geometry

It was in many of these articles as well as his major book of 1832, *Systematische Entwicklung der Abhangigkeit geometrischer Gestalten von Einander* (Systematic Development of the Dependency of Geometrical Forms on One Another) that Steiner laid the foundations for what would become known as projective geometry. Projective geometry is different from the traditional form of geometry taught by Greek geometer **Euclid of Alexandria** (c. 325–c. 270 B.C.E.; see entry in volume 1). It would be called "projective" because it would study the properties involved in transferring or projecting a three-dimensional object to a two-dimensional surface, such as a painter might do. Another way to describe this new geometry would be that it is the study of shadows cast (or projected) by geometric figures. This marks the beginning of a more complicated and thus more modern form of geometry. The work of Steiner, along with that of French mathematician and engineer Jean Victor Poncelet (pronounced pahns-LAY; 1788–1867), would turn projective geometry into a solid and accepted branch of mathematics. Once its principles became fully understood, it would prove essential to many practical fields involving projection, such as making maps of the Earth's surface, shadows cast by objects, and even motion pictures.

Provides for future students

The self-taught Steiner never married and was said to have sometimes been a difficult person. His personality was described as blunt and even abrasive. He was said to have a fiery temper when he felt strongly about something, and he was also known to behave crudely at times, which would turn people away. He always had very liberal political views and never seemed to forget his very humble origins. Over the years, Steiner was able to accumulate a relatively large amount of money. His lecturing proved very popular and most of his savings came from this source. Upon his death in 1863, some of Steiner's estate provided funds for the establishment of a mathematics prize at the Berlin Academy. Besides leaving a large sum of money to his relatives, he also left money to the school in his home village of Utzenstorf to establish prizes for students who excelled in mathematics. Over the course of a long career, he received many high honors and awards, but he never forgot the value of education nor the memory of the lack of it in his own youth.

For More Information

Abbott, David, ed. *The Biographical Dictionary of Scientists: Mathematicians.* New York: Peter Bedrick Books, 1984.

Burckhardt, Johann Jakob. "Jakob Steiner." In *Biographical Dictionary of Mathematicians.* New York: Charles Scribner's Sons, 1991, pp. 2318–27.

Jakob Steiner

Jakob Steiner

O'Connor, J. J., and E. F. Robertson. "Jakob Steiner." *The Mac-Tutor History of Mathematics Archive.* School of Mathematics and Statistics, University of St. Andrews, Scotland. http://www.groups.dcs.st-andrews.ac.uk/~history/Mathematicians/ Steiner.html (accessed June 30, 2002).

Daniel Gabriel Fahrenheit

Born May 24, 1686, in Danzig (now Gdansk), Poland
Died September 16, 1736, in the Hague, Netherlands
German Dutch physicist

Anders Celsius

Born November 27, 1701, in Uppsala, Sweden
Died April 25, 1744, in Uppsala, Sweden
Swedish astronomer

William Thomson (Baron Kelvin of Largs)

Born June 26, 1824, in Belfast, Ireland
Died December 17, 1907, in Netherhall, Ayrshire, Scotland
Scottish mathematician and physicist

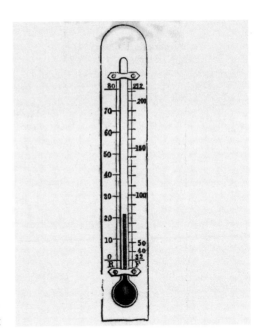

Thermometer Scale Originators

Three men of different countries and living in three different centuries are responsible for the modern thermometer. Today, their names are familiar, as Daniel Gabriel Fahrenheit invented the mercury thermometer and the first useful scale, Anders Celsius changed and improved that scale, and Baron Kelvin took that scale to its ultimate extremes. Mathematics was essential to the discoveries of all three men.

Early thermometers

Despite the considerable advances made in mathematics by the year 1700, there was not yet any sort of reliable, standard temperature scale. Measuring and measurement would always be essential to the progress of science, and physicists, chemists, and even doctors were keenly aware of how imprecise much of their work was without having the proper instrument to tell the temperature of something. Such an instrument is called a thermometer. A ther-

Fahrenheit thermometer.
Reproduced by permission of
Corbis-Bettmann.

mometer measures the temperature of something in a quantitative (of or relating to a certain amount or number) way.

Thermometer Scale Originators

By 1700, people were using a device called a thermoscope to measure temperature. Invented in 1592 by Italian astronomer, physicist, and mathematician **Galileo Galilei** (1564–1642; see entry in volume 1), a thermoscope consisted of a long glass tube with a wide bulb at one end and an opening at the other. When the tube was heated, the air within it would expand and some air was released. The open end was then placed in a dish of water and as the tube cooled, the warm air would contract and draw some of the liquid into the tube. Once it stopped doing this, any later change in air temperature would cause the level of the liquid within the tube to rise or fall. This happens because liquid expands as it is heated and contracts when it is cooled. The problem with this early device was that the tube had no numbers on it (called a scale) in order to measure any change. The first person to put a numerical scale on such an instrument was a contemporary of Galileo's, Italian physician Santorio Santorio (1561–1636). While this could be described as the first thermometer, over the years it would prove wildly unreliable, and in the latter part of the seventeenth century, scientists learned that such an open-ended system would react to air pressure as well as the temperature of the air (in other words, the thermoscope acted as a barometer to measure atmospheric pressure as well).

The thermoscope would remain an object of curiosity for several decades as different experimenters worked toward designing a sealed tube with mercury or wine in it that showed a numerical scale. (Mercury was used because it gave more precise measurements on a scale. Wine, or any type of alcohol, was preferred by many because of its larger expansion, since it expanded more in response to increases in temperature.) As these designs improved, it became apparent to everyone that what was needed next was some sort of absolute scale that everyone could agree on. There were naturally many ideas about how and upon what such a standard should be based, and by 1700 there were some nineteen different temperature scales in common use. Each was based upon a different standard ranging from the melting point of butter to the coolness of the earth inside a below-ground cellar. Given the various scales being used, no one could be sure what anyone else was talking about when they referred to the temperature of something.

Fahrenheit scale

To this disorder came a skilled German-born instrument maker who was living in the Netherlands. Born in Danzig, which is now called Gdansk, Poland, Daniel Gabriel Fahrenheit (pronounced FARE-en-hite) perfected what would become the modern thermometer. As a fifteen-year-old, Fahrenheit was sent to Amsterdam when his wealthy parents, Daniel Fahrenheit and Concordia Schumann, died suddenly. There, he was to study business, but Fahrenheit became acquainted and then fascinated by the fast-growing business of making scientific instruments. When he was twenty-one, Fahrenheit began visiting instrument makers in Germany and Denmark, learning their skills and building his own instruments. By the time he settled back in Amsterdam in 1717, he was ready to set himself up as a maker of fine scientific instruments. Three years before, he had already begun designing and building his own thermometers, using mercury as the liquid inside the tube. Mercury was an ideal liquid to use: When warmed, it expanded quite a bit but always in a uniform way; it stayed liquid over a wide range of temperatures so it would not freeze easily; and its silvery appearance made it easy to see and read. One thing Fahrenheit did to ensure the use of mercury in thermometers was to invent a new method for cleaning mercury so that it would not stick to the inside walls of the glass tube as it rose and fell.

Fahrenheit's most significant contribution, however, was his development of the standard thermometric scale that now bears his name. Fahrenheit knew that any scale needed "fixed points," and many had been suggested by others over the years. For his first low "fixed point," which he called 0 degrees, he chose the point at which water mixed with salt would freeze. Toward the other end of the scale, there were three other normals or fixed points—96 degrees (later adjusted to 98.6) for body temperature, and 32 degrees for freezing regular water. Finally, after performing many experiments on boiling water, he found that its boiling point was constant at 212 degrees. In other words, the temperature never went higher after reaching a certain point (212 degrees), no matter how much more the water boiled. He also discovered that the boiling point could change as the atmospheric pressure changed. This meant that water would boil at a different temperature when high above sea level compared to water boiled at sea level. After Fahrenheit reported his inventions and discoveries to the Royal Society of

London in 1724, they were published as five brief papers, and by 1726, the Fahrenheit temperature scale was on its way to being known and used by every major scientist throughout Europe.

In producing a standard temperature scale that everyone could accept (and an instrument that worked well), Fahrenheit raised the level of scientific precision to a new level and contributed significantly to the future progress of experimental physics and the practice of medicine. England and eventually the rest of the English-speaking world adopted the Fahrenheit scale immediately. Fahrenheit, who never married, died on September 16, 1736, in the Netherlands.

Celsius scale

Although the Fahrenheit scale was used in the English-speaking world, the greater part of the civilized world would eventually use a different scale invented by Anders Celsius two decades after Fahrenheit. Unlike Fahrenheit, Swedish astronomer Anders Celsius (pronounced SELL-see-us) came from a prominent scientific family. Born in Uppsala, Sweden, his father was a University of Uppsala professor of astronomy and mathematics. Celsius studied at Uppsala, and, in 1730, like his father, became a professor of astronomy there. After taking part in an expedition to Lapland (a region that stretches from Russia through Finland) for several years, he was placed in charge of the first observatory built in Sweden.

Celsius is best known today, however, for his 1742 invention of the hundred-point temperature scale, which came to be called the "Celsius scale." Unlike Fahrenheit, Celsius based his scale on two fixed points: the boiling and melting points of water (snow or ice). While the use of the boiling point of water as a temperature calibration (gauge) was not new, Celsius chose to use the melting point rather than the freezing point of water as his second calibration. He obtained this temperature point after performing many experiments placing a thermometer in thawing snow. Between these two fixed points he placed "one hundred steps," thereby making it a *centigrade* (Latin for hundred steps) scale. Celsius first assigned the boiling point at 0 degrees and the melting point at 100 degrees, but soon reversed this to what it is today. In the inevitable competition between Fahrenheit and Celsius temperature scales, Celsius eventually won as it was adopted by the scien-

Swedish astronomer
Anders Celsius.
*Reproduced by permission of
Archive Photos, Inc.*

tific community and used in almost all scientific work. Celsius
died of tuberculosis (a lung disease) at the age of forty-two.

In 1948, most of the non-English-speaking parts of the world
(which make up by far the majority of countries) decided to adopt
the hundred-point scale and agreed to call it the Celsius scale
(instead of the popular Centigrade). By the late 1960s and early
1970s, many of the English-speaking countries (though not the
United States) were also using the Celsius scale.

Thermometer Scale Originators

Kelvin scale

The scientist now known as Lord Kelvin was born in Belfast, Ireland, as William Thomson. As the son of James Thomson, a professor of both engineering and mathematics, he was educated at home before entering the University of Glasgow at the age of eleven. Young Thomson was a prodigy (pronounced PRAH-dih-gee; a young person of extraordinary talent) who attended his father's lectures at the age of eight. At Glasgow, he finished second in his class in mathematics despite his young age, and even published a paper while in his teens. After graduating from Cambridge University in 1845, he went to Paris for postgraduate work and to meet the best scientists in France. He soon accepted the chair of natural philosophy (science) at Cambridge, and remained there for the rest of his career. After a lifetime of great achievement as a mathematician, physicist, and engineer, and becoming universally recognized as the premier scientific mind of the nineteenth century, he was given the title, Sir William Thomson, Baron Kelvin of Largs, in 1892.

As a young man, Thomson was interested in the concepts of heat and energy, and explored ultra-high and ultra-low temperatures of gases. Because of his pioneering work in this emerging field, he proposed that a new temperature scale be devised that measures the ultimate extremes of hot and cold. Thomson then developed the idea of an "absolute" scale of temperature after he made a discovery of great explanatory power. It was known that for every degree Celsius below zero a gas was cooled, its volume would decrease by 1/276. This meant in principle that at –276 degrees, the volume of the gas would be zero. No one could prove this nor even state why this was so, until Thomson explained that when the temperature of a gas is reduced, so is the energy level of the atoms. As the atoms move less, they take up less room, thereby decreasing the volume. At –276 degrees centigrade, the energy of the atoms reaches zero. There, they stop moving and taking up space, and their temperature cannot be lowered any further. Thomson called this absolute zero since it was true for all substances. He then applied this concept to his invention of the absolute scale of temperature. This scale essentially drops the Celsius or centigrade scale by 276 degrees, so that zero and absolute zero coincide. With Thomson's scale, the boiling point of water is 373 degrees. Thomson called his scale the absolute scale, but after

Thermometer Scale Originators

Scottish mathematician and
physicist Lord Kelvin.
*Reproduced by permission of the
Corbis Corporation.*

his death it was renamed the Kelvin scale. Physicists and chemists found the Kelvin scale easier to use and more applicable than the Celsius scale since it represents the last word in thermal scales as it measures the ultimate extremes of hot and cold.

Kelvin's scientific curiosity and ability was very broad, and he studied subjects as diverse as the age of the earth, the phenomenon of heat, and the nature of gases. He also had a great deal to do with the success of the transatlantic telegraph cable which eventually connected England and the United States by an undersea cable in 1858. Kelvin studied the capacity of a cable to carry an electrical current and invented several improvements in design without which the telegraph cable would have been useless. He received a knighthood for this achievement in 1866. Kelvin lived to be eighty-three, but in his later years, the person who began as a child prodigy and went on to great adult achievements, became a man seemingly bewildered by new things. Therefore, the same Kelvin who had put forth so many revolutionary theories and who had introduced Alexander Graham Bell's telephone to England, became a person who reacted against the new, modern physics at the turn of the century, and who bitterly opposed the notion that such unseen things as radioactive atoms could give off energy when they disintegrated.

Today, the name of the man of each thermometer is perpetuated by the constant use of his particular system and no one scale or system has eliminated the use of any other scale. In some ways, each has found at least a niche. Although the Fahrenheit scale was the first widely used temperature scale, it is now only used in the United States to measure temperatures at or near the surface of the Earth. The Celsius scale is used in most of the rest of the world to measure air temperatures, and even the United States uses it to measure upper air temperatures. The Kelvin scale is used by scientists to measure astronomical temperatures. All three scales are related to one another by what is called the "triple point of water." This is the temperature at which liquid water, water vapor, and ice can all coexist at the same time. This triple point occurs at 32.02 degrees Fahrenheit, 0.01 degrees Celsius, and 273.16 degrees Kelvin.

For More Information

"Anders Celsius (1701–1744)." *Uppsala Astronomical Observatory.* http://www.astro.uu.se/history/Celsius_eng.html (accessed June 30, 2002).

Buchwald, Jed Z. "Sir William Thomson." In *Dictionary of Scientific Biography.* Edited by Charles Coulston Gillispie. New York: Charles Scribner's Sons, 1980.

"Fahrenheit, Daniel Gabriel." In *The Biographical Dictionary of Scientists.* Edited by Roy Porter. New York: Oxford University Press, 1994, p. 222.

Gough, J. B. "Daniel Gabriel Fahrenheit." In *Dictionary of Scientific Biography.* Edited by Charles Coulston Gillispie. New York: Charles Scribner's Sons, 1980, pp. 516–18.

Lindroth, Sten. "Anders Celsius." In *Dictionary of Scientific Biography.* Edited by Charles Coulston Gillispie. New York: Charles Scribner's Sons, 1980, pp. 173–74.

O'Connor, J. J., and E. F. Robertson. "William Thomson (Lord Kelvin)." *The MacTutor History of Mathematics Archive.* School of Mathematics and Statistics, University of St. Andrews, Scotland. http://www.groups.dcs.st-andrews.ac. uk/~history/Mathematicians/Thomson.html (accessed June 30, 2002).

Thermometer
Scale Originators

Born 1540
Fontenay-le-Comte, Poitou, France

Died 1603
Paris, France

French algebraist

François Viète

François Viète.
Courtesy of the Library of Congress.

As one of the greatest amateur mathematicians in the history of mathematics, François Viète played a major role in the development of modern **algebra** (see entry in volume 1) by devising algebraic notation. He also introduced algebraic terms that are still in use today, and contributed new techniques of solving very complex equations.

Began a legal career

François Viète (pronounced VYEHT) was born in western France about thirty-five miles from the coast. His father, Etienne Viète, was a lawyer, and his mother, Marguerite Dupont, came from a prominent family. After receiving his early education from the monks at a local Franciscan monastery in his hometown of Fontenay-le-Comte, Viète was sent to the University of Poitiers where he studied law as his father had done. He also studied mathematics, which he always enjoyed. After graduating with a law degree in 1560, he returned to Fontenay where he worked as an attorney. In 1564, Viète decided to change his career and accepted a position as tutor to the daughter of a wealthy and upper-class matron, Antoinette d'Aubeterre. He also agreed to supervise the young woman's education and to serve as private secretary to d'Aubeterre. When d'Aubeterre's husband died, the family moved to La Rochelle, and Viète went with them until the daughter married in 1570. It was during these years of service to the family that Viète found he had enough free time to indulge his strong interest in mathematics.

Service to the Crown

After his student married, Viète's main occupation became his service in royal courts. After moving to Paris in 1570, he somehow caught the attention of King Charles IX (1550–1574) of France, and in 1573, the king appointed him counselor (financial adviser) to Parliament (like the American Congress) in Brittany (one of France's historic provinces located on the northwest coast). He would hold this position until 1580 when a promotion allowed him to return to Paris. He then served as counselor to Charles IX and his successor, Henry III (1551–1589), but in 1584, he was forced from the royal court because he was a Huguenot (pronounced HEW-guh-not; a French Protestant).

Over twenty years earlier, the French wars of religion (1562–98) had begun between French Catholics and French Protestants. By 1584, the Catholics had gained enough control that he was banished from the court (because he was a known Huguenot) and forced by his political enemies to go into exile. For the next five years, he lived in the coastal town of Beauvoir-sur-Mer, and it was during these years that he would do his best mathematical work since he had no formal duties to distract him. By 1589, France was at war with Spain, and Viète was recalled to the court by King Henry, whose parliament was now set up in Tours, France.

Following the assassination of Henry III that same year, Viète went to work for his successor, Henry IV (1553–1610), who used Viète to decode Spanish messages that had been intercepted. Viète's success at codebreaking was such that the Spanish king accused the French of using black magic against his country. Viète remained in service to the court until illness forced him to retire in 1602. During his life at court, Viète was often called by the Latin version of his French name, Franciscus Vieta.

François Viète

François Viète

Mathematical contributions

During his entire life, Viète really had only two extended periods of time when he could conduct any sort of mathematical research and writing. The first was when he was a tutor, and the second was when he was in exile. Although Viète was never a professional mathematician, he sometimes lectured and occasionally published his work. His ties to the royal court allowed his work to be printed by the royal printer. In 1579, his first important book, *Canon mathematicus seu ad triangula cum appendibus* (Mathematical Laws Applied to Triangles), was published, and in 1591 his most influential, *In artem analyticam Isagoge* (Introduction to the Analytical Arts), appeared. It was in this work that Viète introduced the first systematic use of symbolic algebraic notation. This means that he developed the use of letters to represent quantities, both known and unknown. In this work, Viète demonstrated the value of symbols by using the plus and minus signs for operations, vowels for unknown quantities (called variables), and consonants for known quantities (called parameters). Today, a variation of this system is used, as unknowns are commonly represented by $x, y,$ and $z,$ and known quantities by $a, b, c, d,$ and so on. His book demonstrated the power and simplicity of algebra, and helped it develop further. It is also perhaps one of the earliest identifiable "modern" algebra textbooks. Viète also wrote books on trigonometry and **geometry** (see entry in volume 1), and worked on the calculation of **pi** (see entry in volume 2).

Viète played a significant role in the calendar controversies of his time and wrote a criticism of the Copernican theory (which argued correctly that the Earth revolves around the Sun). Viète's work was also studied carefully by French geometer, algebraist and philosopher **René Descartes** (1596–1650; see entry in volume 1) as well as by English mathematician and physicist **Isaac Newton** (1643–1727; see entry in volume 2), and thus contributed heavily to the development of modern mathematics.

There is disagreement as to the date of Viète's death. Some say it was February 13, 1603, while others argue it was December 13, 1603. It is known that he was married twice: to Barbe Cothereau and after her death, to Juliette Leclerc. He had at least one child. There is little available that conveys what Viète was like as a person. However, given that he was only able to devote roughly one-seventh of his life to mathematics, Viète must surely be considered one of the great "amateurs" of mathematics.

For More Information

"Biography of François Viète." *The Art of Algebra from al-Khwarizmi to Viète: A Study in the Natural Selection of Ideas.* University of Virginia, Science and Engineering Libraries. http://www.lib.virginia.edu/science/parshall/viete.html (accessed July 2, 2002).

Busard, H. L. "François Viète." In *Biographical Dictionary of Mathematicians.* New York: Charles Scribner's Sons, 1991, pp. 2512–19.

Franceschetti, Donald R., ed. *Biographical Encyclopedia of Mathematicians.* New York: Marshall Cavendish, 1999.

O'Connor, J. J., and E. F. Robertson. "François Viète." *The MacTutor History of Mathematics Archive.* School of Mathematics and Statistics, University of St. Andrews, Scotland. http://www.groups.dcs.st-andrews.ac.uk/~history/Mathematicians/Viete.html (accessed July 2, 2002).

François Viète

Born November 27, 1923
Chicago, Illinois

African American mathematical physicist

J. Ernest Wilkins Jr.

J. Ernest Wilkins.
Reproduced by permission of
AP/Wide World Photos.

A pioneer in nuclear reactor design, J. Ernest Wilkins Jr. has had a diverse career in government, industry, and academia. As the youngest student ever admitted to the University of Chicago, he became the seventh African American to earn a Ph.D. in mathematics and the second to be named to the National Academy of Engineering.

Nationwide attention as a teenager

Jesse Ernest Wilkins Jr. was born in Chicago, Illinois, the son of J. Ernest Wilkins Sr. and Lucille Beatrice Robinson. The senior Wilkins was a prominent lawyer who was an assistant secretary of labor in the administration of President Dwight D. Eisenhower (1890–1969). Wilkins's mother was a schoolteacher with a master's degree. Wilkins also had two brothers. As a youngster, Wilkins was so bright that he gained admission to college at the age of thirteen. When the University of Chicago accepted him to study mathematics, he became the youngest student ever admitted to that institution. Three and a half years later, he received his bachelor's degree in mathematics, and graduated with an academic record so outstanding that he was ranked in the top ten in the

Putnam Competition, a well-known undergraduate mathematics competition. A year later, he earned his master's degree, and in 1942, at the age of nineteen, he became the seventh African American to obtain a Ph.D. in mathematics. Because he was so young, his achievement attracted nationwide attention. The fact that he came from a minority group seemed to make his doctoral degree in mathematics even more remarkable, and soon Wilkins was described in national newspapers as "the Negro genius."

Begins diverse career

After receiving his Ph.D., Wilkins knew he was most interested in the applications of mathematics, and decided that this would become his specialty. That same year, he received a Rosenwald Scholarship and became a fellow (a visiting member) at Princeton's Institute for Advanced Study. In 1943, Wilkins began his teaching career as an instructor of mathematics at the Tuskegee Institute, an historic school for blacks founded in 1881 by African American educator and reformer Booker T. Washington (1856–1915). Wilkins remained there only one year as World War II (1939–45) created a national demand for people with his mathematical knowledge. So in 1944, Wilkins joined the U.S. government's secret effort to build the first atomic bomb. He returned to the University of Chicago where he became an associate physicist at its Metallurgical Laboratory, which was part of the Manhattan Project (the code name given to the American effort to build the first atomic bomb). He remained there until 1946 and was promoted to physicist.

After the war, Wilkins found the greatest opportunity in American industry rather than its academic institutions. In 1946, he accepted a position as mathematician at the American Optical Company in Buffalo, New York. It was during this time that Wilkins experienced racism when, in 1947, he received an invitation from the American Mathematics Society (AMS) to its southeastern region meeting. The letter informed him, however, that the society had arranged for him to stay with a "nice colored family" who would also supply his meals for him. He declined the invitation and has not attended an AMS meeting in the southeast since then.

In 1950, Wilkins left Buffalo to become senior mathematician at the Nuclear Development Corporation of America, which later became the United Nuclear Corporation in White Plains, New

J. Ernest
Wilkins Jr.

J. Ernest Wilkins Jr.

York. He would remain there for fifteen years and eventually become manager of research and development. During his years with this corporation, Wilkins obtained both a bachelor's degree (1957) and a master's of mechanical engineering degree (1960) from New York University. From 1960 to 1970, Wilkins worked at the General Atomic Company in San Diego, California, rising in positions and responsibility. In 1970, he joined Howard University in Washington, D.C., as distinguished professor of applied mathematical physics, and remained there until 1977, during which he also served as a visiting scientist for one year at the Argonne National Laboratory.

In 1977, Wilkins went to EG&G Idaho in Idaho Falls, Idaho, where he rose to vice president and deputy general manager. After leaving there in 1984, he became a fellow at the Argonne National Laboratory for a year. This institution, part of the U.S. Department of Energy in Argonne, Illinois, is involved in the research and development of peaceful uses of nuclear energy. When he retired in 1985, he remained an active consultant for Argonne. In 1990, Wilkins joined Clark Atlanta University where he became distinguished professor of applied mathematics and mathematical physics.

Contributions to gamma ray research

Ever since he was involved with nuclear energy on the Manhattan Project, Wilkins applied his mathematics to nuclear engineering, his primary area of interest. One of his major accomplishments was the development of radiation shielding against gamma radiation. Gamma rays or radiation is emitted (given off) during the natural electron decay of the Sun and other nuclear sources. This type of radiation can be destructive to living tissue and must be shielded. Wilkins is best known for his work with Herbert Goldstein (1922–) on gamma-ray penetration, the results of which are used for the design of nuclear reactors and for radiation shielding. Wilkins developed mathematical models by which the amount of gamma radiation that is absorbed by a given material can be calculated. This technique eventually became widely used for research in space and for other nuclear science projects.

During his career, Wilkins published roughly one hundred papers and reports on pure (theoretical) and applied mathematics, nuclear engineering, and optics (the science of light). He also wrote several papers on reactor operation and design and heat transfer. During his career, he served on advisory committees on scientific and engineering education, and is recognized for playing a significant role in helping Howard University establish a Ph.D. program in mathematics. Among his many awards and honors, perhaps the most significant is his election to the National Academy of Engineering (1976). He was also awarded the Outstanding Civilian Service Medal by the U.S. Army in 1980. Wilkins has been married twice: to Gloria Stewart and to Maxine Grundy Malone, each of whom is deceased. Wilkins and his first wife had two children, Sharon and J. Ernest III.

The breadth and range of Wilkins's mathematical achievements, especially as applied to the fields of physics and nuclear energy, show that he was a child prodigy who lived up to the great promise that he demonstrated as a youngster.

For More Information

Houston, Johnny L. "J. Ernest Wilkins, Jr." *SUMMA Website, The Mathematical Association of America.* http://www.maa.org/summa/archive/WilkinsJ.htm (accessed July 3, 2002).

"J. Ernest Wilkins, Jr." *The Faces of Science: African Americans in the Sciences.* http://www.princeton.edu/~mcbrown/display/wilkins.html (accessed July 3, 2002).

O'Connor, J. J., and E. F. Robertson. "Jesse Ernest Wilkins Jr." *The MacTutor History of Mathematics Archive.* School of Mathematics and Statistics, University of St. Andrews, Scotland. http://www.groups.dcs.st-andrews.ac.uk/~history/Mathematicians/Wilkins_Ernest.html (accessed July 3, 2002).

Williams, Scott W. "J. Ernest Wilkins, Jr." *Mathematicians of the African Diaspora.* The Mathematics Department of the State University of New York at Buffalo. http://www.math.buffalo.edu/mad/PEEPS/wilkns_jearnest.html (accessed July 3, 2002).

J. Ernest
Wilkins Jr.

Selected Bibliography

General sources

Asimov, Isaac. *Realm of Numbers.* Boston: Houghton Mifflin, 1959.

Ball, W. W. Rouse. *A Short Account of the History of Mathematics.* New York: Dover Publications, 1960.

Bergamini, David. *Mathematics.* Alexandria, VA: Time-Life Books, 1980.

Borman, Jami Lynne. *Computer Dictionary for Kids—and Their Parents.* Hauppauge, NY: Barron's Educational Series, 1995.

Boyer, Carl B., and Uta C. Merzbach. *A History of Mathematics.* New York: John Wiley & Sons, 1989.

Bunt, Lucas N. H., et al. *The Historical Roots of Elementary Mathematics.* Englewood Cliffs, NJ: Prentice Hall, 1976.

Burton, David M. *Burton's History of Mathematics.* Dubuque, IA: Wm. C. Brown Publishers, 1995.

Cajori, F. *A History of Mathematics.* New York: Chelsea, 1985.

Selected Bibliography

Dictionary of Mathematics Terms. New York: Barron's Educational Series, Inc., 1987.

Duren, Peter, ed. *A Century of Mathematics in America.* 3 vols. Providence, RI: American Mathematical Society, 1989.

Eves, Howard. *An Introduction to the History of Mathematics.* Philadelphia: Saunders College Publishing, 1990.

Flegg, Graham. *Numbers: Their History and Meaning.* New York: Schocken Books, 1983.

Friedberg, Richard. *An Adventurer's Guide to Number Theory.* New York: Dover Publications, 1994.

Green, Gordon W., Jr. *Helping Your Child to Learn Math.* New York: Citadel Press, 1995.

Green, Judy, and Jeanne Laduke. *A Century of Mathematics in America.* Providence, RI: American Mathematical Society, 1989.

Groza, Vivian Shaw. *A Survey of Mathematics: Elementary Concepts and Their Historical Development.* New York: Holt, Rinehart and Winston, 1968.

Heath, T. L. *A History of Greek Mathematics.* New York: Dover Publications, 1981.

Heddens, James W. and William R. Speer. *Today's Mathematics: Concepts and Methods in Elementary School Mathematics.* Upper Saddle River, NJ: Merrill, 1997.

Hirschi, L. Edwin. *Building Mathematics Concepts in Grades Kindergarten Through Eight.* Scranton, PA: International Textbook Co., 1970.

Hoffman, Paul. *Archimedes' Revenge: The Joys and Perils of Mathematics.* New York: Ballantine, 1989.

Hogben, Lancelot T. *Mathematics in the Making.* London: Galahad Books, 1974.

Humez, Alexander, et al. *Zero to Lazy Eight: The Romance of Numbers.* New York: Simon & Schuster, 1993.

Immergut, Brita. *Arithmetic and Algebra—Again.* New York: McGraw-Hill, 1994.

Julius, Edward H. *Arithmetricks: 50 Easy Ways to Add, Subtract, Multiply, and Divide Without a Calculator.* New York: John Wiley and Sons, 1995.

Katz, Victor J. *A History of Mathematics: An Introduction.* New York: HarperCollins College Publishers, 1993.

Kline, Morris. *Mathematics for the Nonmathematician.* New York: Dover Publications, 1985.

Kline, Morris. *Mathematics in Western Culture.* New York: Oxford University Press, 1953.

Miles, Thomas J., and Douglas W. Nance. *Mathematics: One of the Liberal Arts.* Pacific Grove, CA: Brooks/Cole Publishing Co., 1997.

Miller, Charles D., et al. *Mathematical Ideas.* Reading, MA: Addison-Wesley, 1997.

Moffatt, Michael. *The Ages of Mathematics: The Origins.* Garden City, NY: Doubleday & Company, 1977.

Rogers, James T. *The Pantheon Story of Mathematics for Young People.* New York: Pantheon Books, 1966.

Slavin, Steve. *All the Math You'll Ever Need.* New York: John Wiley and Sons, 1989.

Smith, David Eugene. *Number Stories of Long Ago.* Detroit: Gale Research, 1973.

Smith, David Eugene, and Yoshio Mikami. *A History of Japanese Mathematics.* Chicago: The Open Court Publishing Company, 1914.

Smith, Karl J. *Mathematics: Its Power and Utility.* Pacific Grove, CA: Brooks/Cole, 1997.

Stillwell, John. *Mathematics and Its History.* New York: Springer-Verlag, 1989.

Temple, George. *100 Years of Mathematics.* New York: Springer-Verlag, 1981.

Selected Bibliography

West, Beverly Henderson, et al. *The Prentice-Hall Encyclopedia of Mathematics.* Englewood Cliffs, NJ: Prentice-Hall, 1982.

Wheeler, Ruric E. *Modern Mathematics.* Pacific Grove, CA: Brooks/Cole Publishing, 1995.

Wheeler, Ruric E., and Ed R. Wheeler. *Modern Mathematics for Elementary School Teachers.* Pacific Grove, CA: Brooks/Cole Publishing, 1995.

Wulforst, Harry. *Breakthrough to the Computer Age.* New York: Charles Scribner's Sons, 1982.

General biographical sources

Abbott, David, ed. *The Biographical Dictionary of Scientists: Mathematicians.* New York: Peter Bedrick Books, 1986.

Albers, Donald J., and G. L. Alexanderson, eds. *Mathematical People: Profiles and Interviews.* Boston: Birkhauser, 1985.

Albers, Donald J., Gerald L. Alexanderson, and Constance Reid. *More Mathematical People.* New York: Harcourt, 1991.

Alec, Margaret. *Hypatia's Heritage: A History of Women in Science from Antiquity through the Nineteenth Century.* Boston: Beacon Press, 1986.

Asimov, Isaac. *Asimov's Biographical Encyclopedia of Science and Technology.* Garden City, NY: Doubleday & Company, 1982.

Bell, Eric T. *Men of Mathematics.* New York: Simon and Schuster, 1986.

Biographical Dictionary of Mathematicians. New York: Charles Scribner's Sons, 1991.

Cortada, James W. *Historical Dictionary of Data Processing: Biographies.* New York: Greenwood Press, 1987.

Daintith, John, et al. *Biographical Encyclopedia of Scientists.* London: Institute of Physics Publishing, 1994.

Dunham, W. *The Mathematical Universe: An Alphabetical Journey through the Great Proofs, Problems, and Personalities.* New York: John Wiley & Sons, 1994.

Elliott, Clark A. *Biographical Dictionary of American Science: The Seventeenth Through the Nineteenth Centuries.* Westport, CT: Greenwood Press, 1979.

Gillispie, Charles C., ed. *Dictionary of Scientific Biography.* New York: Charles Scribner's Sons, 1990.

Grinstein, Louise S., and Paul J. Campbell, eds. *Women of Mathematics: A Biobibliographic Sourcebook.* New York: Greenwood Press, 1987.

Haber, Louis. *Black Pioneers of Science and Invention.* New York: Harcourt, Brace & World, 1970.

Henderson, Harry. *Modern Mathematicians.* New York: Facts on File, 1996.

Hollingdale, Stuart. *Makers of Mathematics.* London: Penguin Books, 1989.

Hudson, Wade, and Valerie Wilson Wesley. *Afro-Bets Book of Black Heroes From A to Z: An Introduction to Important Black Achievers for Young Readers.* East Orange, NJ: Just Us Books, 1997.

It, Kiyosi, ed. *Encyclopedia Dictionary of Mathematics.* Cambridge, MA: MIT Press, 1987.

McGraw-Hill Modern Scientists and Engineers. New York: McGraw-Hill, 1980.

McMurray, Emily J., ed. *Notable Twentieth-Century Scientists.* Detroit: Gale, 1995.

Metcalf, Doris Hunter. *Portraits of Exceptional African American Scientists.* Carthage, IL: Good Apple, 1994.

Millar, David, Ian Millar, John Millar, and Margaret Millar. *The Cambridge Dictionary of Scientists.* Cambridge, England: Cambridge University Press, 1996.

Morgan, Bryan. *Men and Discoveries in Mathematics.* London: John Murray Publishers, 1972.

Morrow, Charlene, and Teri Perl, eds. *Notable Women in Mathematics: A Biographical Dictionary.* Westport, CT: Greenwood Press, 1998.

Selected Bibliography

Muir, Jane. *Of Men and Numbers: The Story of the Great Mathematicians*. New York: Dover Publications, 1996.

Ogilvie, Marilyn Bailey. *Women in Science: Antiquity through the Nineteenth Century*. Cambridge, MA: MIT Press, 1986.

Osen, Lynn M. *Women in Mathematics*. Cambridge, MA: The MIT Press, 1974.

Pappas, Theoni. *Mathematical Scandals*. San Carlos, CA: Wide World Publishing/Tetra, 1997.

Perl, Teri. *Math Equals: Biographies of Women Mathematicians*. Menlo Park, CA: Addison-Wesley Publishing Company, 1978.

Porter, Roy, ed. *The Biographical Dictionary of Scientists*. New York: Oxford University Press, 1994.

Potter, Joan, and Constance Claytor. *African Americans Who Were First: Illustrated with Photographs*. New York: Cobblehill Books, 1997.

Reimer, Luetta, and Wilbert Reimer. *Mathematicians Are People, Too: Stories from the Lives of Great Mathematicians*. Palo Alto, CA: Dale Seymour Publications, 1995.

Ritchie, David. *The Computer Pioneers*. New York: Simon and Schuster, 1986.

Shasha, Dennis E. *Out of Their Minds: The Lives and Discoveries of 15 Great Computer Scientists*. New York: Copernicus, 1998.

Simmons, George F. *Calculus Gems: Brief Lives and Memorable Mathematics*. New York: McGraw-Hill, 1992.

Slater, Robert. *Portraits in Silicon*. Cambridge, MA: The MIT Press, 1989.

Spencer, Donald D. *Great Men and Women of Computing*. Ormond Beach, FL: Camelot, 1999.

Young, Robyn V., ed. *Notable Mathematicians: From Ancient Times to the Present*. Detroit: Gale Research, 1998.

Internet sites

Readers should be reminded that some Internet sources change frequently. All of the following web sites were accessible as of August 17, 2002, but some may have changed addresses or been removed since then.

The Abacus
http://www.ee.ryerson.ca:8080/~elf/abacus/

American Mathematical Society (AMS)
http://e-math.ams.org/

Ask Dr. Math
http://forum.swarthmore.edu/dr.math/

Athena: Earth and Space Science for K-12
http://inspire.ospi.wednet.edu:8001/

Biographies of Women Mathematicians
http://www.agnesscott.edu/lriddle/women/women.htm

Brain Teasers
http://www.eduplace.com/math/brain/

Canadian Mathematical Society
http://camel.cecm.sfu.ca/CMS

Eisenhower National Clearinghouse for Mathematics and Science
http://www.enc.org/

Explorer
http://explorer.scrtec.org/explorer/

Flashcards for Kids
http://www.edu4kids.com/math

Fraction Shapes
http://math.rice.edu/~lanius/Patterns/

Galaxy
http://galaxy.einet.net/galaxy/Science/Mathematics.html

MacTutor History of Mathematics Archive
http://www-history.mcs.st-andrews.ac.uk/history/

Math Forum: Elementary School Student Center
http://forum.swarthmore.edu/students/students.elementary.html

Math Forum: Math Magic!
http://forum.swarthmore.edu/mathmagic/

Math Forum: Women and Mathematics
http://forum.swarthmore.edu/social/math.women.html

Math League Help Topics
http://www.mathleague.com/help/help.htm

Mathematical Association of America
http://www.maa.org/

Mathematical Programming Glossary
http://carbon.cudenver.edu/~hgreenbe/glossary/glossary.html

The Mathematics Archives
http://archives.math.utk.edu/

Mathematics Web Sites Around the World
http://www.math.psu.edu/MathLists/Contents.html

Mathematics WWW Virtual Library
http://euclid.math.fsu.edu/Science/math.html

Measure for Measure
http://www.wolinskyweb.com/measure.htm

Mega Mathematics!
http://www.c3.lanl.gov/mega-math/

Past Notable Women of Computing and Mathematics
http://www.cs.yale.edu/~tap/past-women.html

PlaneMath
http://www.planemath.com

Women in Math Project
http://darkwing.uoregon.edu/~wmnmath/

The Young Mathematicians Network WWW Site
http://www.youngmath.org/

Organizations

American Mathematical Society
201 Charles St.
Providence, RI 02904-2294
Internet site: http://www.ams.org

American Statistical Association
1429 Duke Street
Alexandria, VA 22314-3415
Internet site: http://www.amstat.org

Association for Women in Mathematics
4114 Computer and Space Science Building
University of Maryland
College Park, MD 20742-2461
Internet site: http://www.awm-math.org

Association of Teachers of Mathematics
7 Shaftesbury Street
Derby DE23 8YB England
Internet site: http://www.atm.org.uk/

Institute of Mathematical Statistics
P.O. Box 22718
Beachwood, OH 44122
Internet site: http://www.imstat.org/

Math/Science Interchange
c/o Department of Mathematics
Loyola Marymount University
7900 Loyola Blvd.
Los Angeles, CA 90045

Math/Science Network
Mills College
5000 MacArthur Boulevard
Oakland, CA 94613-1301
Internet site: http://www.expandingyourhorizons.org/

Mathematical Association of America
1529 18th Street N.W.
Washington, DC 20036-1385
Internet site: http://www.maa.org

National Council of Supervisors of Mathematics
P.O. Box 10667
Golden, CO 80401
Internet site: http://forum.swarthmore.edu/ncsm

Selected Bibliography

National Council of Teachers of Mathematics
1906 Association Drive
Reston, VA 20191-1502
Internet site: http://www.nctm.org

School Science and Mathematics Association
400 East 2nd Street
Bloomsburg, PA 17815
Internet site: http://www.ssma.org

Society for Industrial and Applied Mathematics
3600 University City Science Center
Philadelphia, PA 19104-2688
Internet site: http://www.siam.org/nnindex.htm

Women and Mathematics Education
c/o Dorothy Buerk
Mathematics and Computer Science Department
Ithaca College
Ithaca, NY 14850-7284
Internet site: http://www.wme-usa.org/

Index

This is a cumulative index of *Math and Mathematics,* volumes 1–3. *Italic* type indicates volume number; **boldface** indicates main entries and their page numbers; (ill.) indicates photos and illustrations.

Index

Lagrange, Joseph-Louis, *2:* 253–58

Noether, Emmy, *2:* 319–24

noncommutative algebra, *1:* 210

Oresme, Nicole d', *3:* 125–30

Turing, Alan, *2:* 427–32

Viète, François, *3:* 164–67

word origin, *3:* 5

Algebraic notation

Viète, François, *3:* 164–67

ALGOL (Algorithmic Language), *3:* 103

Algorithm, *1:* 8; *2:* 241, 244, 394, 410

word origin, *3:* 5

Algorithmic Language (ALGOL), *3:* 103

al-Khwārizmī. *See* Khwārizmī, al-

Allen, Paul, *3:* 103

Almagest (The Greatest) (Ptolemy), *3:* 138

al-Majisti (Great Work) (Ptolemy), *3:* 138

Analog clock, *2:* 418

Analog computer, *1:* 82

Analysis, *1:* 134

Cantor, Georg, *1:* 53–57

Chang, Sun-Yung Alice, *3:* 43–46

Granville, Evelyn Boyd, *1:* 201–05

Hadamard, Jacques-Salomon, *3:* 77–82

Lambert, Johann, *3:* 105–10

Analytic functions, *3:* 78

Analytic geometry (coordinate graphing), *1:* 87–90, 89 (ill.), 99, 141, 143

Oresme, Nicole d', *3:* 125–130

The Analytical Arts Applied to Solving Algebraic Equations (Harriot), *3:* 87

Analytical engine, *1:* 33, 82; *3:* 113

Analytical Society, *1:* 32

Angle, *2:* 422

Apollonius, *1:* 142, 226, *2:* 237

Applied mathematics

acoustics, *1:* 192

elasticity, *1:* 192

geodesy, *1:* 181

Germain, Sophie, *1:* 189–93

Granville, Evelyn Boyd, *1:* 201–5

Lovelace, Ada, *3:* 111–17

Arab scholars

Greek mathematics, *3:* 5

Arabic numerals, *1:* 148; *2:* 243–44; *3:* 5

Arago, François, *1:* 135

Archimedean screw, *1:* 22

Archimedes of Syracuse, *1:* 21 (ill.), **21–26**, 23 (ill.), 137; *2:* 330, 331 (ill.)

Area, *1:* **27–29**, 78; *2:* 326, 330, 334, 422, 434. *See also* Perimeter

Aristotle, *2:* 282, 285 (ill.)

Arithmetic operations, *1:* 108; *2:* 410

Arithmetica logarithmica (The Arithmetic of Logarithms) (Briggs), *3:* 25

Artificial intelligence, *2:* 392

Turing, Alan, *2:* 431

Wiener, Norbert, *2:* 445–50

Artin, Emil, *1:* 73

Artis Analyticae Praxis ad Aequationes Algebraicas Resolvendas (The Analytical Arts Applied to Solving Algebraic Equations) (Harriot), *3:* 87

Association for Women in Mathematics (AWM), *3:* 73–74

Associative property, *1:* 8, 108; *2:* 294, 410

Astronomy

Banneker, Benjamin, *1:* 37–43

Briggs, Henry, *3:* 23–27

Bürgi, Joost, *1:* 33–37

Galileo, *1:* 163–69

Gauss, Carl Friedrich, *1:* 177–82

Index

British Association for the Advancement of Science, *1:* 35

Brouncker, William, *3:* 28 (ill.); *3:* **28–32**

Brown, Robert, *1:* 113

Browne, Marjorie Lee, *1:* 201, 203, 203 (ill.)

Brownian motion, *1:* 113; *2:* 447

Bug (computer term), *1:* 215

Bureau of Ordnance Computation Project, *1:* 215

Bürgi, Joost, *3:* 33 (ill.), **33–37**

Bush, Vannevar, *2:* 390

Byron, Augusta Ada. *See* Lovelace, Ada

Byron, Lord (George Gordon), *3:* 111, 116

C

Calculating machines, *1:* 82; *2:* 302. *See also* Computer

Calculator, digital, *3:* 59, 133, 133 (ill.)

Calculus, *1:* 145; *2:* 316; *3:* 135
    Babbage, Charles, *1:* 31–36
    first comprehensive textbook, *1:* 13
    Leibniz, Gottfried, *2:* 259–65
    Newton, Isaac, *2:* 311–18
    Seki Kōwa, *2:* 385

Calculus of variations, *2:* 253, 254

Caliph al-Ma'mūn, *2:* 242

Canon mathematicus seu ad triangula cum appendibus (Mathematical Laws Applied to Triangles) (Viète), *3:* 166

Cantor, Georg, *1:* 53 (ill.), **53–57,** *3:* 55

Carbon absorption, *2:* 418

Cardano, Girolamo, *1:* 59 (ill.), **59–64**

Carr, G. S., *2:* 366

Carroll, Lewis, *2:* 284, 284 (ill.)

Cartan, Elie-Joseph, *1:* 73, 73 (ill.)

Cartography, Ptolemaic, *3:* 140–41

Cataldi, Pietro Antonio, *1:* 167

Cauchy, Augustin-Louis, *1:* **65–69,** 65 (ill.), 172

Cayley, Arthur, *3:* 38 (ill.), **38–41**

Celsius, Anders, *3:* **158–59,** 159 (ill.)

Census Bureau, *3:* 62

Ceres, *1:* 179

Chang, Sun-Yung Alice, *3:* **43–46**

Charles I (king of England), *3:* 12, 28, 29

Charles IX (king of France), *3:* 165

Chaucer, Geoffrey, *1:* 118

Chebyshev, Pafnuti L., *2:* 344

Chebyshev's theorem, *1:* 120

Chern, Shiing-Shen, *1:* 71 (ill.), **71–75**

Chios, Greece, *3:* 89

Chladni, E. F. F., *1:* 191

Chladni figures, *1:* 191

Chord, *1:* 78

Chromatic polynomials, *3:* 10

Church, Alonzo, *2:* 429

Circadian rhythm, *2:* 418

Circle, *1:* **77–80,** 79 (ill.), 80 (ill.)
    squaring, *3:* 92–93

Circulus, *1:* 77

Circumference, *1:* 24, 78, 184; *2:* 268, 269, 326, 330, 376. *See also* Perimeter
    of Earth, *1:* 129–30

Classical mechanics, *3:* 96–97

Claytor, William, *3:* 49

Clepsydra, *2:* 418

Clocks, *1:* 165, 222; *2:* 418, 420; *3:* 37

COBOL (programming language), *1:* 213, 217

Colburn, Zerah, *1:* 208

COLOSSUS (computer), *2:* 430

Colson, John, *1:* 14

Columbus, Christopher, *3:* 140–41

Common (Briggsian) logarithms, *3:* 25

Communications theory
    Shannon, Claude E., *2:* 389–92

Commutative property, *1:* 8, 108, *2:* 294, 410

Index

Index

woman professor of mathematics on a university faculty, *1:* 11

First-degree equations, *1:* 20

Fixed points, in thermometric scale, *3:* 157, 158

FLOW-MATIC (programming language), *1:* 217

Fontana, Niccolò, *1:* 61, 62, 62 (ill.)

Foot (measurement), *2:* 267

Formula, *1:* 28; *2:* 340

FORTRAN (FORmula TRANslation), *3:* 103

Fourier, Joseph, *1:* 153 (ill.), **153–57**

Fourier's theorem, *1:* 153, 155

Fourth dimension, *1:* 210; *2:* 382

Fractal geometry, *2:* 287, 290

Fraction, *1:* 150, **159–62;** *2:* 278, 403

    continued, *3:* 30

    irrational numbers, *3:* 54

    unit, *1:* 160; *3:* 4

Fraction bar symbol, *1:* 160

Frederick the Great (king of Prussia), *3:* 107, 109

Frederick William IV (king of Prussia), *3:* 97

Freezing point of water, in thermometric scale, *3:* 157

French Revolution, *3:* 121–123

Frey, Gerhard, *2:* 453

Fundamenta nova theorae functionum ellipticarum (New Foundations of the Theory of Elliptic Functions) (Jacobi), *3:* 96

G

Galilei, Galileo. *See* Galileo

Galileo, *1:* 163 (ill.), **163–69,** 164 (ill.), 166 (ill.), 168, 179, 221, 222; *3:* 156

Galley method (division), *1:* 108

Galois, Évariste, *1:* 171 (ill.), **171–75**

Galton, Francis, *2:* 398, 400 (ill.)

Game theory, *1:* 47

economics, *2:* 309

Neumann, John von, *2:* 305–310

Gamma rays (gamma radiation), *3:* 170

"Garbage in, garbage out," *3:* 115

Gases, in Kelvin scale, *3:* 160

Gates, Bill, *3:* 103

Gauss, Carl Friedrich, *1:* 177 (ill.), **177–82,** 180 (ill.); *2:* 272, 380, 381–82; *3:* 18, 20 (ill.), 20–21, 52

General Electric Corporation, *3:* 102

General equation of the fifth degree, *1:* 3

General theory of relativity, *1:* 117; *2:* 319–24, 382

Genetics, *2:* 344, 373

Geocentric theories, *3:* 129, 139, 140 (ill.), 141

Geodesy, *1:* 181

Geographike hyphegesis (Guide to Geography) (Ptolemy), *3:* 139, 141

Geography

    Ptolemy, Claudius, *3:* 137–41

Geometric diagrams, *1:* 127 (ill.)

Geometric progression, *3:* 4

Geometrical Lectures (Barrow), *3:* 15

Geometrical optics, *2:* 240

Geometry, *1:* **183–87**

    Agnesi, Maria, *1:* 11–15

    analytic (coordinate graphing), *1:* 87–90, 89 (ill.), 99, 141, 143

    Archimedes of Syracuse, *1:* 21–26

    Barrow, Isaac, *3:* 11–15

    Bolyai, János, *3:* 17–22

    Briggs, Henry, *3:* 23–27

    Brouncker, William, *3:* 28–32

    Cayley, Arthur, *3:* 38–41

    Chang, Sun-Yung Alice, *3:* 43–46

    Chern, Shiing-Shen, *1:* 71–75

    coordinate geometry, *1:* 103

    Descartes, René, *1:* 99–105

O

Index

P

Pangeometry, *2:* 274

Pappus of Alexandria, *1:* 142, 220

Papyrus, *1:* 18, 28; *2:* 422; *3:* 1
  Ahmes (Rhind), *3:* 1–4

Parabolas, *2:* 352

Parallel postulate, *2:* 274; *3:* 19, 107

Parallelogram, *1:* 28; *2:* 326, 362

Parameters, *3:* 166

Parchment, *2:* 340

Partitioning, *2:* 368

Pascal, Blaise, *1:* 82, 141, 144; *2:* 220, 261, 345, 345 (ill.); *3:* 131 (ill.), **131–36**

Pascal's principle, *3:* 134

Pearson, Karl, *2:* 398, 401 (ill.)

Pendulums, *1:* 165

Pentagon, *2:* 326, 337 (ill.)

Pepys, Samuel, *3:* 30–31, 31 (ill.)

Percent, *2:* **371–74**

Percent symbol (%), *2:* 373

Percy, Henry, *3:* 85–86

Perfect tautochronism, *1:* 222

Perimeter, *1:* 28, 29, 78; *2:* 268, 269, **325–27**, 330, 334, 422, 434. *See also* Circumference

Perpendicular, *1:* 88; *2:* 422

Perspective, *1:* 184, 185

Pestalozzi, Johann Heinrich, *3:* 151 (ill.), 151–52

Philosophy
  Descartes, René, *1:* 99–105
  Hypatia of Alexandria, *1:* 225–29
  Leibniz, Gottfried, *2:* 259–65
  Pythagoras of Samos, *2:* 349–53
  Thales of Miletus, *2:* 413–16

Photoelectric effect, *1:* 113

Photometria (Photometry) (Lambert), *3:* 108

Physics, mathematical, *1:* 166, 189, 192
  12th century, *1:* 147
  17th century, *1:* 141
  18th century, *2:* 253
  Babylonia, *2:* 329, 356, 393

discrete mathematics, *1:* 119

Egypt, *1:* 160; *2:* 397

Einstein, Albert, *1:* 111–18

Euler, Leonhard, *1:* 135

Galileo, *1:* 163–69

Germain, Sophie, *1:* 189–93

Greece, *1:* 127

Hadamard, Jacques-Salomon, *3:* 81

Huygens, Christiaan, *1:* 219–23

Jacobi, Carl, *3:* 94–98

Japan, *2:* 385

in literature, *1:* 118

logical method, *1:* 128

music, *2:* 351

Neumann, John von, *2:* 305–10

Newton, Isaac, *2:* 311–18

Pascal, Blaise, *3:* 134

quaternions, *1:* 210

Riemann, Bernhard, *2:* 379–83

Stevin, Simon, *2:* 403–7

Wilkins, J. Ernest, Jr., *3:* 168–71

Pi, *1:* 79; *2:* 326, **329–32**, 376
  Archimedes of Syracuse, *1:* 24
  Babylonia, *2:* 330
  Brouncker, William, *3:* 28–32
  China, *2:* 330
  Egypt, *2:* 330
  Lambert, Johann, *3:* 105–10

Pi symbol (π), *2:* 331

Pisano, Leonardo. *See* Fibonacci, Leonardo Pisano

Place value (positional notation), *1:* 8, 96; *2:* 410, 442

Planck, Max, *1:* 113

Plane geometry, *1:* 128

Planetary motion, laws of, *2:* 235, 239

Planetary orbits, *2:* 238 (ill.)

Plato, *2:* 353

Plus symbol (+), *1:* 8; *2:* 295; *3:* 166

Poe, Edgar Allan, *1:* 118

Poincaré, Jules Henri, *1:* 55

Index

Transatlantic telegraph cable, *3:* 161–62

Transfinite numbers, *1:* 55

Trapezoid, *1:* 28; *2:* 362

Triangle, *2:* 356, **421–26,** 423 (ill.), 425 (ill.). *See also* Polygon; Pythagorean theorem

Triangulation, *2:* 422, 423

Trigonometry, *2:* 278, 299–303, 358, 421; *3:* 139

Triple point of water, *3:* 162

Troy weight, *2:* 438

Turin Academy of Sciences, *2:* 254

Turing, Alan, *2:* 391, 427 (ill.), **427–32,** 428 (ill.)

Turing machine, *2:* 429

Two-term unit fraction, *3:* 4

U

Unit fraction, *1:* 160, *3:* 4

Universal Automatic Computer (UNIVAC), *3:* 62

U.S. Air Force Chapel (Colorado Springs, CO), *2:* 424 (ill.)

V

Vacuum, *3:* 134

Vacuum tube, *3:* 60

Variable, *1:* 103; *2:* 387; *3:* 166

Variations, calculus of, *2:* 253, 254

Veblen, Oswald, *1:* 74

Verne, Jules, *1:* 118

Vertex, *2:* 334, 422

Vesalius, Andreas, *1:* 61

Vibration patterns, *1:* 191

Viète, François, *1:* 142; *3:* 164 (ill.), **164–67**

Viviani, Vincenzo, *3:* 13

Volume, *1:* 25; *2:* 330, **433–36**

von Neumann, John. *See* Neumann, John von

Von Salis, Peter, *1:* 106

W

Wallis, John, *1:* 109, 110 (ill.); *3:* 30

Washington, D.C., *1:* 39

Water
freezing and boiling points, *1:* 58; *3:* 157

Water clocks, *2:* 418

Water snail, *1:* 22

Weber, Wilhelm, *1:* 181; *2:* 381

Weierstrass, Karl T., *2:* 249

Weight, *2:* 435, **437–39**

Weyl, Hermann, *1:* 74

Whole numbers, *2:* 232, **441–44**

Widmann, Johannes, *1:* 9; *2:* 295, 411

Wiener, Norbert, *2:* 445 (ill.), **445–50,** 448 (ill.)

Wiles, Andrew J., *1:* 144; *2:* 451 (ill.), **451–56,** 454 (ill.)

Wilhelm IV (duke of Hesse-Kassel), *3:* 34

Wilkins, J. Ernest, Jr., *3:* 168 (ill.), **168–71**

Williams, Lloyd, *3:* 49

Witch of Agnesi, *1:* 14

Wolf Prize in Mathematics, *1:* 75, 121

Women mathematicians
Agnesi, Maria, *1:* 11–15
Bari, Ruth Aaronson, *3:* 7–10
Breteuil, Gabrielle-Émilie le Tonnelier de, *2:* 263
Browne, Marjorie Lee, *1:* 203
Chang, Sun-Yung Alice, *3:* 43–46
Fasenmyer, Sister Mary Celine, *3:* 127
Germain, Sophie, *1:* 189–93
Granville, Evelyn Boyd, *1:* 201–5
Gray, Mary, *3:* 71–75
Hoover, Erna Schneider, *3:* 69
Hopper, Grace, *1:* 213–18
Hypatia of Alexandria, *1:* 225–29
Kovalevskaya, Sofya, *2:* 247–51
Lovelace, Ada, *3:* 111–17
Noether, Emmy, *2:* 319–24

Robinson, Julia Bowman, *3:* 143–48
Woodard, Dudley, *3:* 49

X

X-axis, *1:* 88

Y

Y-axis, *1:* 88

Yang, *1:* 80
Yang, Chen-Ning, *1:* 75
Yard (measurement), *2:* 268
Yau, Shing-Tung, *1:* 75
Year, *2:* 418
Yenri, *2:* 387
Yin, *1:* 80

Z

Zero, *1:* 149; *2:* 232, 243

Index